T0282507

CAMBRIDGE LIBRARY COLLECTION

Books of enduring scholarly value

Technology

The focus of this series is engineering, broadly construed. It covers technological innovation from a range of periods and cultures, but centres on the technological achievements of the industrial era in the West, particularly in the nineteenth century, as understood by their contemporaries. Infrastructure is one major focus, covering the building of railways and canals, bridges and tunnels, land drainage, the laying of submarine cables, and the construction of docks and lighthouses. Other key topics include developments in industrial and manufacturing fields such as mining technology, the production of iron and steel, the use of steam power, and chemical processes such as photography and textile dyes.

The Principles of Bridges

Though raised in Newcastle's coal-mining community, Charles Hutton (1737–1823) went on to make his mark as a teacher and mathematician. A fellow of the Royal Society (and recipient of the Copley medal), he carried out research into the convergence of series, ballistics, and the density of the earth. After flooding destroyed several bridges across the Tyne in November 1771, he began to study the design of bridges, and published this mathematical treatment in 1772. It demonstrates the ideal properties of arches and piers, with due consideration given to the force of water flowing against these structures. Hutton's practical observations also enhance a section that provides definitions of relevant terms. Not merely a solution to the demands of transport and trade, a well-designed bridge, in Hutton's eyes, stands as a structure of elegance and beauty.

Cambridge University Press has long been a pioneer in the reissuing of out-of-print titles from its own backlist, producing digital reprints of books that are still sought after by scholars and students but could not be reprinted economically using traditional technology. The Cambridge Library Collection extends this activity to a wider range of books which are still of importance to researchers and professionals, either for the source material they contain, or as landmarks in the history of their academic discipline.

Drawing from the world-renowned collections in the Cambridge University Library and other partner libraries, and guided by the advice of experts in each subject area, Cambridge University Press is using state-of-the-art scanning machines in its own Printing House to capture the content of each book selected for inclusion. The files are processed to give a consistently clear, crisp image, and the books finished to the high quality standard for which the Press is recognised around the world. The latest print-on-demand technology ensures that the books will remain available indefinitely, and that orders for single or multiple copies can quickly be supplied.

The Cambridge Library Collection brings back to life books of enduring scholarly value (including out-of-copyright works originally issued by other publishers) across a wide range of disciplines in the humanities and social sciences and in science and technology.

The Principles of Bridges

Containing the Mathematical Demonstrations of the Properties of the Arches, the Thickness of the Piers, the Force of the Water against Them, &c.

CHARLES HUTTON

CAMBRIDGE
UNIVERSITY PRESS

University Printing House, Cambridge, CB2 8BS, United Kingdom

Cambridge University Press is part of the University of Cambridge.
It furthers the University's mission by disseminating knowledge in the pursuit of education, learning and research at the highest international levels of excellence.

www.cambridge.org
Information on this title: www.cambridge.org/9781108070492

This edition first published 1772
This digitally printed version 2014

ISBN 978-1-108-07049-2 Paperback

THE

PRINCIPLES

OF

BRIDGES:

CONTAINING THE

MATHEMATICAL DEMONSTRATIONS

OF

The PROPERTIES of the ARCHES, the THICKNESS of the PIERS, the FORCE of the WATER againſt them, &c.

TOGETHER WITH

PRACTICAL OBSERVATIONS and DIRECTIONS drawn from the whole.

By CHA. HUTTON,
MATHEMATICIAN.

NEWCASTLE:

Printed by T. SAINT; and ſold by J. WILKIE in St. Paul's Church Yard, and H. TURPIN, in St. John's Street, London; and by KINCAID and CREECH, Edinburgh, 1772.

PREFACE.

A Large and elegant bridge, forming a way over a broad and rapid river, is justly esteemed one of the noblest pieces of mechanism that man is capable of performing. And the usefulness of an art which, at the same time that it connects distant shores by a way over the deep and rapid waters, also allows those waters and their navigation to pass smooth and uninterrupted, renders all probable attempts to advance the theory or practice of it, highly deserving the encouragement of the publie.

This little book is offered as an attempt towards the perfection of the theory of this art, in which the properties, dimensions, proportions, and other relations of the various parts of a bridge, are strictly demonstrated, and clearly illustrated by various examples. It is divided into five sections: the 1st treats on the projects of bridges, containing a regular detail of the various circumstances and considerations that are cognizable in such projects: The 2d treats on arches, demonstrating their various properties, with the relations between their intrados and extrados, and clearly distinguishes the most preferable curves to be used in a bridge; the first two or three propositions being instituted after the manner of two or three done by Mr. Emerson in his Fluxions and Mechanics: The 3d section treats on the piers, demonstrating their thickness necessary for supporting any kind of an arch, springing at any height, and that both when part of the pier is supposed to be immersed in water, and when otherwise: The 4th demonstrates the force of the water against the end or face of the pier, considered as of different forms; with the best form for dividing the stream, &c. and to it is added a table shewing the several heights of the fall of the water under the arches, arising from its velocity and the obstruction of the piers; as it was composed by Tho. Wright, Esq; of Auckland, in the county of Durham, who informs me it is part of a work on which he has spent much time, and with which he intends to favour the public: And the 5th and last section contains a dictionary of the most material terms peculiar to the subject; in in which many practical observations and directions are given, which could not be so regularly nor properly introduced into the former sections. The whole, it is presumed, containing full directions for constituting and adapting to one another, the several essential parts of a bridge, so as to make it the strongest, and the most convenient, both for the passage over and under it, that the situation and other circumstances will possibly admit: not indeed for the actual methods of disposing the stones, making of mortar, or the external ornaments, &c. those things I do not descend to, but leave to the discretion of the practical architect, as being no part of the plan of my undertaking; and for the same reason also I have given no views of bridges, but only prints of such parts or figures as are necessary in explaining the elementary parts of the subject.

As my profession is not that of an architect, very probably I should never have turned my thoughts to this subject, so as to address the public upon it, had it not been from the occasion of an accident in that part of the country in which I reside, viz. the fall of Newcastle and other bridges on the river Tyne on the 17th of november 1771, occasioned by a high flood which rose about 9 feet higher at Newcastle than the usual spring tides do.——And this occasion having furnished me with many opportunities of hearing and seeing very absurd things advanced on the subject in general, I thought the demonstrations of the relations of the essential parts of a bridge, would not be unacceptable to those architects and others who may be capable of perceiving the force of them, and whose ignorance may not have prejudiced them against things which they do not understand.

In the 4th section there is one thing forgotten to be remarked, viz. That in determining the best form of the end of the pier to be a right-lined triangle, the water was supposed to strike every part of it with the same velocity: had the variably increased velocity been used, the form of the ends would come out a little curved; but as the increase of the velocity in the best bridges is very small, the difference in them is quite imperceptable.

THE

THE

PRINCIPLES

OF

STONE BRIDGES.

SECTION I.

Of the Projects of Bridges, with the Design, Estimate, &c.

WHEN a bridge is deemed necessary to be built over a river, the first consideration is the place of it; or what particular situation will contain a maximum of the advantages over the disadvantages.

In agitating this most important question, every circumstance, certain and probable, attending or likely to attend the bridge, should be separately, minutely, and impartially stated and examined; and the advantage or disadvantage of it rated at a value proportioned to it: then the difference between the whole advantages and diſadvantages, will be the neat value of that particular ſituation for which the calculation is made. And by doing the ſame for any other ſituations, all their neat values will be found, and of conſequence the moſt preferable ſituation among them.——Or, in a competition between two places, if each one's advantage over the other be eſtimated or valued in every circumſtance attending them, the ſums of their advantages will ſhew whether of them is the better. And the ſame being done for this and a third, and ſo on, the beſt ſituation of all will be obtained.

In this eſtimation, a great number of particulars muſt be included; and nothing omitted that can be found to make a part of the conſideration.

Among theſe, the ſituation of the town or place for the convenience of which the bridge is chiefly to be made, will naturally produce a particular of the firſt conſequence; and a great many others ought to be ſacrificed to it. If poſſible, the bridge ſhould be placed where there can conveniently be opened and made paſſages or ſtreets from the ends of it in every direction, and eſpecially one as nearly in the direction of the bridge itſelf as poſſible, tending towards the body of the town, without narrows or crooked windings, and eaſily communicating with the chief ſtreets, thoroughfares, &c.——And here every

every person, in judging of this, should divest himself of all partial regards or attachments whatever; think and determine for the good of the whole only, and for posterity as well as the present.

The banks or declivities towards the river are also of particular concern, as they affect the conveniency of the passage to and from the bridge, or determine the height of it, upon which in a great measure depends the expence.

The breadth of the river, the navigation upon it, and the quantity of water to be passed, or the velocity and depth of the stream, form also considerations of great moment; as they determine the bridge to be higher or lower, longer or shorter. However, in most cases, a wide part of the river ought rather to be chosen than a narrow one, especially if it is subject to great tides or floods; for, the increased velocity of the stream in the narrow part, being again augmented by the farther contraction of the breadth, by the piers of the bridge, will both incommode the navigation through the arches, and undermine the piers and endanger the whole bridge.

The nature of the bed of the river is also of great concern, it having a great influence on the expence; as upon it, and the depth and velocity of the ſtream, depend the manner of laying the foundations, and building the piers.

Theſe are the chief and capital articles of conſideration, and which will branch themſelves out into other dependent ones, and ſo lead to the required eſtimate of the whole.

HAVING reſolved on the place, the next conſiderations are the form, the eſtimate of the expence, and the manner of execution.

With reſpect to the form; ſtrength, utility, and beauty ought to be regarded and united; the chief part of which lies in the arches. The form of the arches will depend on their height and ſpan; and the height on that of the water, the navigation, and the adjacent banks. They ought to be made ſo high, as that they may eaſily tranſmit the water at its greateſt height either from tides or floods; and their height and figure ought alſo to be ſuch as will eaſily allow of a convenient paſſage of the craft through them. This and the diſpoſition of it above, ſo as to render the paſſage over it alſo convenient, make up its utility.—— Having fixed the heights of the arches, their ſpans are ſtill neceſſary for determining their figure. Their ſpans will be known by dividing the whole breadth of the river into a convenient number of arches and piers, allowing at leaſt the neceſſary thickneſs of the

the piers out of the whole. In fixing on the number of arches, take always an odd number, and rather take few and large ones than many and ſmaller, if convenient: For thus you will have not only fewer foundations and piers to make, with fewer arches and centers, which will produce great ſavings in the expence, but the arches themſelves will alſo require much leſs materials and workmanſhip, and allow of more and better paſſage for the water and craft through them; and will appear at the ſame time more noble and beautiful, eſpecially if conſtructed in elliptical, or in cycloidal forms: for the truth of which it may be ſufficient to refer to that noble and elegant bridge lately built at Blackfriars, London, by Mr. Mylne. And here I can't help remarking that the Gentleman who, a few years ſince in a pamphlet on the Principles of Bridges, cenſured Mr. Mylne and Mr. Muller concerning elliptic arches, has very much expoſed himſelf, and abſurdly criticiſes them through his own want of mathematical knowledge, which he ſomewhere in the ſame pamphlet affects to deſpiſe. He brings to my mind an expreſſion of (I think) Mr. Henry Fielding ſomewhere in his works, That a perſon does not ſpeak the worſe on a ſubject for knowing ſomething about it. I do not however make this remark through any particular diſreſpect for this Gentleman, concerning whom I know nothing farther, any more than I do about the other two Gentlemen, but only to prevent

prevent others from being prejudiced and misled by the authority of his *ipse dixit.*—If the top of the bridge be a streight horizontal line, let the arches be made all of a size; if it be a little lower at the ends than the middle, the arches must proportionally decrease from the middle towards the ends; but if higher at the ends than the middle, let them increase towards the ends. A choice of the most convenient arches is to be made from the 4th and 5th propositions, where their several properties, &c. are demonstrated and pointed out: Among them, the elliptic, cycloidal, and equilibrial arch in prop. 5, will generally claim the preference, both on account of their strength, beauty, and cheapness or saving in materials and labour: Other particulars also concerning them may be seen under the word ARCH in the Dictionary in the last section. And as the choice of the arch is of so great moment, let no person, either through ignorance or indolence, prefer a worse arch because it may seem to him easier to construct; for he would very ill deserve the name or employment of an Architect, who is incapable of rendering the exact construction of these curves easy and familiar to himself; but if, by chance, a Bridge-builder should be employed who is incapable of doing that, he ought at least to be endowed with such a share of honesty as to procure some person to go through the calculations which he cannot make for himself.

Next

Next find what thickness at the keystone or top will be necessary for the arches. For which see the word KEYSTONE in the Dictionary in the last section.

Having thus obtained all the parts of the arches, with the height of the piers, the necessary thickness of the piers themselves are next to be computed by prop. 10.

This done, the chief and material requisites are found; the elevation and plans of the design can then be drawn, and the calculations of the expence from thence made, including the foundations, with such ornamental or accidental appendages as shall be thought fit; which I shall leave to the discretion of the Practical Architect, as being no part of the plan of my undertaking, together with the practical methods of carrying the design into execution. I shall however, in the Dictionary in the last section, not only describe the terms, parts, machines, &c. but also speak of their dimensions, properties, and any thing else material belonging to them; and to which therefore I from hence refer for more explicit information in each particular article, as well as to these immediately following propositions, in which the theory of the arches, piers, &c. are fully and strictly demonstrated.

SECTION II.

Of the Arches.

PROPOSITION I.

LET there be any number of lines AB, BC, CD, DE, *&c. all in the ſame vertical plane, connected together and moveable about the joints or angles* A, B, C, D, E, F; *the two extreme points* A *and* F *being fixed: It is required to find the proportions of the weights to be laid upon the angles* B, C, D, *&c. ſo that the whole may remain in equilibrium.*

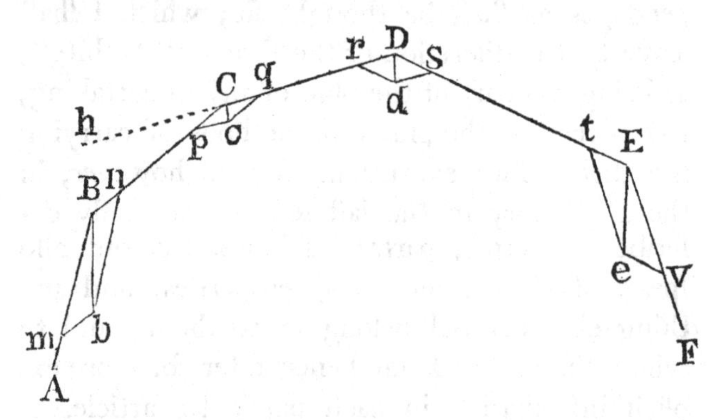

Solution.

FROM the ſeveral angles having drawn the lines Bb, Cc, Dd, &c. perpendicular to the horizon; about them, as diagonals, conſtitute paral-

parallelograms ſuch, that thoſe ſides of each two that are upon the ſame one of the given lines, may be equal to each other; viz. having made one parallelogram m n, take C p = B n, and form the parallelogram p q; then take D r = C q, and make the parallelogram r s; and take E t = D s, and form the parallelogram t v; and ſo on: Then the ſaid vertical diagonals B b, C c, D d, E e, &c. of thoſe parallelograms, will be proportional to the weights, as required.

Demonſtration.

By the reſolution of forces, each of the weights or forces B b, C c, D d, &c. in the diagonals of the parallelograms, is equal to, and may be reſolved into two forces expreſſed by two adjacent ſides of the parallelogram; viz. the force B b will be reſolved into the two forces B m, B n, and in thoſe directions; the force C c into the two forces C p, C q, and in thoſe directions; the force D d into the two forces D r, D s, and in thoſe directions; and ſo on: Then, ſince two forces that are equal, and in oppoſite directions, do mutually balance each other; therefore the ſeveral pairs of forces B n and C p, C q and D r, D s and E t, &c. being equal and oppoſite, by the conſtruction, do mutually deſtroy or balance each other; and the extreme forces B m, E v, are balanced by the oppoſite reſiſtances of the fixed points A, F. Wherefore there is no force to change the poſition of any one of the lines, and conſequently they will all remain in equilibrium. *Q.E.D.*

Corollary.

HENCE, if one of the weights and the poſitions of all the lines be given, all the other weights may be found.

PROPOSITION II.

IF any number of lines, that are connected together and moveable about the points of connection, be kept in equilibrium by weights laid upon the angles, as in the laſt propoſition: Then will the weight on any angle C *be univerſally as* $\frac{\text{ſine of the } \angle \text{BCD}}{\text{s.} \angle \text{BCc} \times \text{s.} \angle \text{cCD}}$; *that is, directly as the ſine of that angle, and reciprocally as the ſines of the two parts or angles into which that angle is divided by a line drawn through it perpendicular to the horizon.*

Demonſtration.

BY the laſt propoſition the weights are as Bb, Cc, Dd, &c. when Bn = pC, Cq = rD, Ds = tE, &c. But, ſince the angle ABb is = the angle Bbn,

Bbn, and the angle BCc = the angle Ccq, &c. as being always the alternate angles made by a line cutting two other parallel lines; also the sine of the $\angle$ ABC = s. $\angle$ Bnb, and s. $\angle$ BCD = s. $\angle$ Cqc, as being supplements one to another; by plane trigonometry we shall have

$$(Bn =) \frac{Bb \times s. \angle ABb}{s. \angle ABC} = (Cp =) \frac{Cc \times s. \angle cCD}{s. \angle BCD},$$

$$(Cq =) \frac{Cc \times s. \angle BCc}{s. \angle BCD} = (Dr =) \frac{Dd \times s. \angle dDE}{s. \angle CDE},$$

$$(Ds =) \frac{Dd \times s. \angle CDd}{s. \angle CDE} = (Et =) \frac{Ee \times s. \angle eEF}{s. \angle DEF},$$

&c.

Hence

$$Bb : Cc :: \frac{s. \angle ABC}{s. \angle ABb} : \frac{s. \angle BCD}{s. \angle cCD},$$

$$Cc : Dd :: \frac{s. \angle BCD}{s. \angle BCc} : \frac{s. \angle CDE}{s. \angle dDE},$$

$$Dd : Ee :: \frac{s. \angle CDE}{s. \angle CDd} : \frac{s. \angle DEF}{s. \angle eEF},$$

&c.

Or, by dividing the latter terms of the first of these proportions each by s. $\angle$ bBC, and then compounding together two of the proportions, then three of them, &c. striking out the common factors, and observing that the s. $\angle$ bBC is = s. $\angle$ BCc, the s. $\angle$ cCD = s. $\angle$ CDd, &c. we shall have

$$Bb : Cc :: \frac{s. \angle ABC}{s. \angle ABb \times s. \angle bBC} : \frac{s. \angle BCD}{s. \angle BCc \times s. \angle cCD},$$

$$Bb : Dd :: \frac{s. \angle ABC}{s. \angle ABb \times s. \angle bBC} : \frac{s. \angle CDE}{s. \angle CDd \times s. \angle dDE},$$

$$Bb : Ee :: \frac{s. \angle ABC}{s. \angle ABb \times s. \angle bBC} : \frac{s. \angle DEF}{s. \angle DEe \times s. \angle eEF},$$

&c.

Q.E.D.

Otherwise.

SINCE Cp or Bn : Bm or nb :: s. $\angle$ Bbn or s. $\angle$ ABb : s. $\angle$ bBC or s. $\angle$ BCc :: $\frac{1}{s. \angle BCc} : \frac{1}{s. \angle ABb}$,

and Cp or qc : Cq or Dr :: s. $\angle$ cCq or s. $\angle$ CDd : s. $\angle$ Ccq or s. $\angle$ BCc :: $\frac{1}{s. \angle BCc} : \frac{1}{s. \angle CDd}$;

it is clear that Cp is as $\frac{1}{s. \angle BCc}$; that is, the forces mB, pC, rD, &c. are reciprocally as the sines of the angles which they make with the vertical line.

And since Cc is $= \frac{Cp \times s. \angle Cpc}{s. \angle Ccp} = \frac{Cp \times s. \angle BCD}{s. \angle cCD}$; therefore any force Cc is as $\frac{s. \angle BCD}{s. \angle cCB \times s. \angle cCD}$. *Q.E.D.*

Corol-

Corollary.

If DC be produced to h; the ſine of the ∠ hCB being = to the ſine of its ſupplement BCD, the weight or force Cc will be as $\frac{s. \angle hCB}{s. \angle BCc \times \angle cCD}$; which three angles together make up two right angles.

PROPO-

PROPOSITION III.

TO find the proportion of the height of the wall above every point of an arch of equilibration: That is, if GHIK *be the top of a wall ſupported by an arch* ABCD *; it is required to find the proportion of the perpendiculars* BH, CI, *&c. ſo that all the parts of the arch may be kept in equilibrium from falling, by the weight or preſſure of the ſuperincumbent wall.*

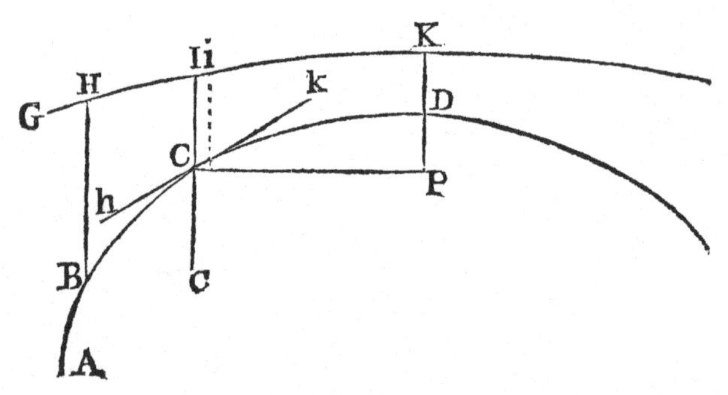

Solution.

THE lines of equilibration in the former propoſitions being imagined to become indefinitely ſmall, they will conſtitute a curve of equilibration, and the weights will preſs upon every point of it, and will be reſpectively equal to the perpendiculars

pendiculars BH, CI, &c. drawn into their reſpective breadths, ſuppoſing them to be indefinitely narrow parallelograms: Alſo the angle hCB will become the angle of contact formed by the tangent and curve, whoſe ſine is equal to the angle itſelf or its meaſure, and the angles cCB and cCD become equal to the angles cCh, cCk, or equal to the angles ICk, ICh, whoſe ſines are equal, becauſe the angles are ſupplements to each other. Theſe values being ſubſtituted in the expreſſion in the corollary to the laſt propoſition, we ſhall have the force Cc or parallelogram Ci as $\frac{\text{the angle } hCB}{\overline{s. \angle hCI}|^{2}}$ or as $\frac{\text{the} \angle kCD}{\overline{s. \angle kCI}|^{2}}$.

Now ſupppoſing theſe narrow parallelograms to ſtand upon indefinitely ſmall equal parts of the arch, their breadths will be directly as the s. $\angle$ kCI and inverſly as radius; hence the parallelogram IC × s. $\angle$ kCI is as $\frac{\text{the} \angle kCD}{\overline{s. \angle kCI}|^{2}}$, and conſequently the altitude IC as $\frac{\text{the} \angle kCD}{\overline{s. \angle kCI}|^{3}}$ or as the $\angle$ kCD × $\overline{\text{ſecant} \angle kCP}|^{3}$; CP being perpendicular to CI, and the radius all along equal to unity.

But the angle of contact kCD is as the curvature of the arch, and that again is inverſly as as the radius of curvature; wherefore IC is as $\frac{1}{R \times \overline{s. \angle kCP}^3}$ or as $\frac{\overline{\text{sec.} \angle kCP}^3}{R}$, putting *R* for the radius of curvature to the point C; that is, the height of the wall above any point, is reciprocally as the radius of curvature and cube of the sine of the angle in which the vertical line cuts the curve in that point, or reciprocally as the radius of curvature and directly as the cube of the secant of the curve's inclination to the horizon.

Corollary 1.

HENCE, if the form of the arch, or nature of the curve ABCD be given, the form of the line GHIK bounding the top of the wall or forming the extrados, may be found so, that ABCD shall be an arch of equilibration, or be in equilibrium in all its parts by the pressure of the wall.

For, since the arch is given, the radius of curvature and position of the tangent at every point of it will be given, and consequently the proportions of the verticals BH, CI, &c. And by assuming one of them, or making it equal to an assigned length, the rest will be found from it; and then the line GHI &c. may be drawn through the extremities of them all.

Corol-

Corollary 2.

AND if the line GHIK, forming the top of the wall be given, the curve of equilibration ABCD may be found. And the manner of finding them both, the one from the other, we ſhall teach in the two following propoſitions.

Corollary 3.

IF the arch ABCD be a circle; the radius of curvature will be conſtant, and the angle kCP always meaſured by the arc DC, ſuppoſing D the vertex of the curve; and then CI will be every-where as the cube of the ſecant of the arc DC.

PROPOSITION IV.

HAVING given the Intrados, to find the Extrados. That is, given the nature or form of an arch, to find the nature of the line forming the top of the ſuperincumbent wall, by whoſe preſſure the arch is kept in equilibrium.

Solution.

LET D be the vertex of the given curve ABCD, and K that of the required line GHIK. Put a = DK, x = AP the abſciſſa, y = PC the ordinate, z = DC the arch, and R = the radius of curvature at the point C.

Now, by the laſt prop. CI is as $\frac{\text{ſec.} \angle \text{kCP}^3}{R}$.

But, by ſimilar triangles, as $\dot{y} : \dot{z} :: 1$ (radius) $: \frac{\dot{z}}{\dot{y}}$ = ſec. $\angle$ kCP; therefore CI is as $\frac{\dot{z}^3}{R\dot{y}^3}$.

Again, in every curve whoſe ordinate is refered to an axis, the radius of curvature R is $= \frac{\dot{z}^3}{\dot{y}\ddot{x} - \dot{x}\ddot{y}}$; wherefore CI will be as $\frac{\dot{y}\ddot{x} - \dot{x}\ddot{y}}{\dot{y}^3}$, or CI $= \frac{\dot{y}\ddot{x} - \dot{x}\ddot{y}}{\dot{y}^3} \times Q$; where Q is a conſtant quantity

whoſe

whofe value will be determined by taking the expreffion for the given perpendicular DK at the vertex of the curve.

Corollary.

HENCE then, as either x or y may be fuppofed to flow uniformly, and confequently either of their fecond fluxions equal to nothing, by ftriking either of the terms out of the numerator of the above value of CI, and then exterminating either of the unknown quantities by the equation of the curve, the value of CI will be obtained; as is done in the following examples.

EXAMPLE I.

To find the extrados of a circular arch.

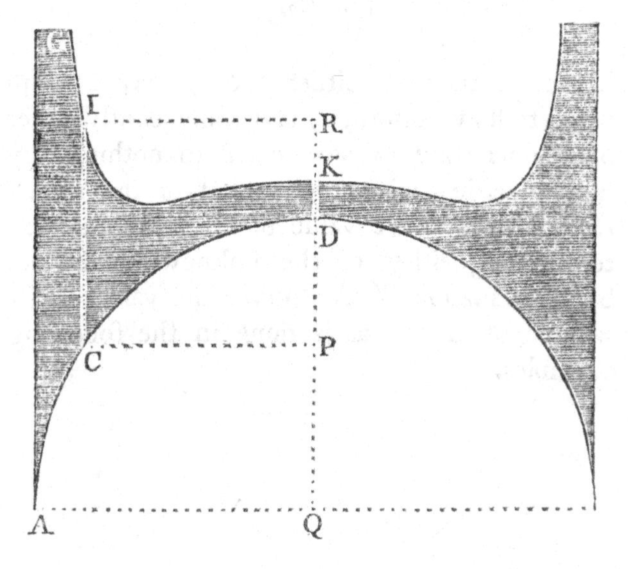

LET Q be the center and D the vertex of the given circular arch, K the vertex of the extrados, and the other lines as in the figure.

Put $a = DK$, $r = AQ = QD =$ the radius, $x = DP$, and $y = PC = RI$.

Then $y = \sqrt{2rx - xx}$, $\dot{y} = \frac{r - x}{\sqrt{2rx - xx}} \times \dot{x}$, and $\ddot{y} = \frac{-r^2 \dot{x}^2}{\overline{2rx - xx}|^{\frac{3}{2}}}$, by making $\ddot{x} = 0$. Hence

CI

$CI = \frac{\dot{y}\ddot{x} - \dot{x}\ddot{y}}{\dot{y}^3} \times Q$ is $= \frac{-\dot{x}\ddot{y}}{\dot{y}^3} \times Q = \frac{r^2 \dot{x}^3}{\overline{2rx - xx}|^{\frac{3}{2}}}$ $\times \frac{\overline{2rr - xx}|^{\frac{3}{2}}}{\overline{r - x}|^3 \times \dot{x}^3} \times Q = \frac{r^2 Q}{\overline{r - x}|^3}$. But, at the vertex x is $= 0$, and then CI is $=$ DR $= a = \frac{r^2 Q}{\overline{r - 0}|^3}$ $= \frac{r^2 Q}{r^3} = \frac{Q}{r}$. Consequently the value of Q is $= ar$. And the general value of CI or $\frac{r^2 Q}{\overline{r - x}|^3}$ is $a \times \overline{\frac{r}{r - x}}|^3 = \frac{DK \times DQ^3}{PQ^3}$.

Otherwise,

By making y constant.

THE notation remaining as before: we have $x = r - \sqrt{r^2 - y^2}$, $\dot{x} = \frac{y\dot{y}}{\sqrt{r^2 - y^2}}$, and $\ddot{x} =$ $\frac{r^2 \dot{y}^2}{\overline{rr - yy}|^{\frac{3}{2}}}$. Hence CI or $\frac{\dot{y}\ddot{x} - \dot{x}\ddot{y}}{\dot{y}^3} \times Q$ becomes $\frac{\ddot{x}}{\dot{y}^2} \times Q = \frac{r^2 Q}{\overline{rr - yy}|^{\frac{3}{2}}}$. This when $y = 0$, gives $a = \frac{Q}{r}$, and $Q = ar$ as before. And conse-

quently

quently CI or $\frac{r^2Q}{\overline{rr-yy}|^{\frac{3}{2}}}$ is $= a \times \overline{\frac{r}{\sqrt{r^2-y^2}}}|^3$ $= \frac{DK \times DQ^3}{PQ^3}$ as before.

Hence the equation to the curve KI is $v =$ $(KR = ax + x - IC =)\ a + x - \frac{ar^3}{\overline{r-x}|^3}$ or $=$ $a + r - \sqrt{r^2 - y^2} - \frac{ar^3}{\overline{rr-yy}|^{\frac{3}{2}}}$.

Corollary 1.

HENCE KIG is a curve running up an infinite height towards G, the perpendicular AG being an asymptote to it: And the curve is accurately as represented in the figure, when the thickness DK at the top is 1–15th of the span.

Corollary 2.

BUT the curve KIG is quite inconvenient for the form of the extrados of any bridge; however a streight horizontal line IK might be used instead of it, if the materials of which the arch is built, could be

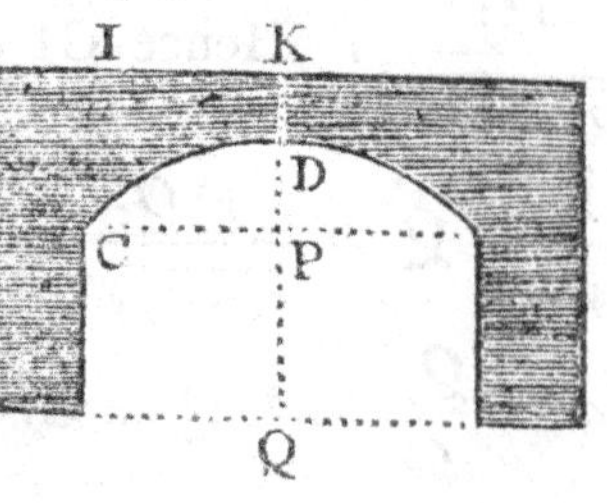

be so chosen, as that they might increase in their specific gravity from DK towards CI, continually as the cube of the secant of the arch from D. And this again perhaps would be quite impracticable: But if a circular arch and a right line at the top were *necessarily* required, the proportion of DK to the radius DQ may be found so as the arch may be *nearly* in equilibrium thus:

When KI is a right line, then KR in the figure to the example, must be nothing; or rather when the curve crosses the horizontal line, then KR is equal to nothing; put its value then, as found above, equal to o, and we shall have $\frac{ar^3}{\overline{rr-yy}|^{\frac{3}{2}}} - a - r + \sqrt{r^2 - y^2} = 0$, and from this equation, by assuming one of the quantities, a, y, the corresponding value of the other may be found for the point where the curve crosses the horizontal line; so from hence the general value of a is $\left(\frac{r - \sqrt{r^2 - y^2}}{r^3 - \overline{\sqrt{r^2 - y^2}}|^3} \times \overline{rr - yy}|^{\frac{3}{2}} = \right) \frac{\overline{rr - yy}|^{\frac{3}{2}}}{r^2 + r\sqrt{r^2 - y^2} + r^2 - y^2} = \frac{PQ^3 = v^3}{r^2 + rv + v^2} = \frac{\overline{r - x}|^3}{3r^2 - 3rx + x^2}$. Now this value of a or DK evidently becomes = o when the arch consists of the whole semi-circle; but when

when the arch is leſs than the ſemicircle, a will have a finite value, and between 60 and 120 degrees many arches of equilibration of a certain thickneſs at top may be found. Thus, if the half arch DC contain 30 degrees; then its ſine y or PC is $= \frac{1}{2}r$; which being ſubſtituted for it in the above general value of a, we have $a = \frac{7\sqrt{3}-6}{37} \times \frac{3}{2}r$, or $= \frac{1}{4}r$ extremely near; that is, DK is $= \frac{1}{4}$ of DQ or $\frac{1}{4}$ of 2PC the ſpan when the curve cuts the horizontal line directly above the point in the circle which anſwers to 30 degrees. And if DC were an arch of 45 degrees; then $y = r\sqrt{\frac{1}{2}}$, and $a = \frac{3\sqrt{2}-2}{14} \times r = \frac{16r}{100}$, or $\frac{1}{9}$ of the ſpan nearly. Alſo, if DC were 60 degrees; then $y = r\sqrt{\frac{3}{4}}$, and $a = \frac{1}{14}$th of $r = \frac{7r}{100}$, or $\frac{1}{16}$ of the ſpan nearly.—— So that in each of theſe caſes the points C and D would be in equilibrium; but then about the middle parts between D and C, or rather nearer to D than to C, the materials ſhould be a little lighter than at D and C, and the exact proportion in which their gravity ſhould be diminiſhed, might eaſily be found by calculation; ſo in the firſt caſe, in particular, the ſpecific gravity of the materials in the middle of the arch between D and C, that is at 15 degrees from D, ſhould be to that at D or C, as 278 to 284, which is but

but a very inconsiderable decrease, and may be very well neglected.——In the first two cases, the thickness at the top would be too much; but in the latter one, when the whole arch is 120 degrees, the thickness is just about that which the best architects now allow; and in greater arches the thickness would become too little. So that an arch of nearly about 120 degrees, is the only part of a circle that can be used with any degree of propriety.

EXAMPLE 2.

TO determine the extrados of an elliptical arch of equilibration.

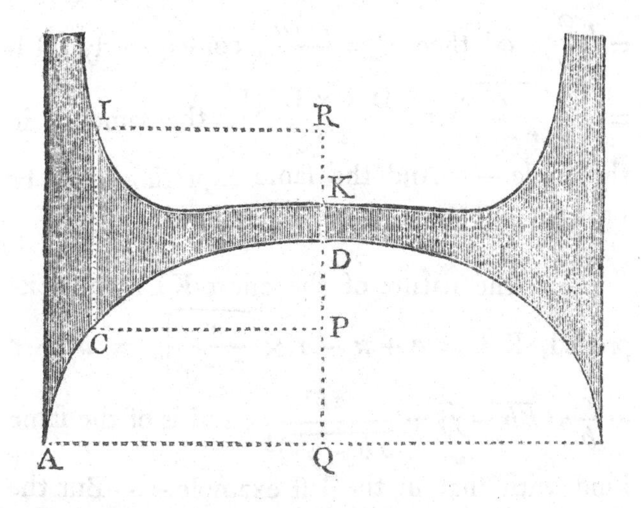

SUPPOSE the curve in the above figure to be a semi-ellipse, with either the longer or shorter axe horizontal; and let h denote the horizontal femi-axe AQ, and r the vertical one DQ, and all the other letters as in the laft example.

Then, by the nature of the ellipfe, $r : h :: \sqrt{2rx - xx} : y = \frac{h}{r}\sqrt{2rx - xx}$; hence $\dot{y} = \frac{h\dot{x}}{r} \times \frac{r - x}{\sqrt{2rx - xx}}$, and $\ddot{y} = \frac{-hr\dot{x}^2}{\overline{2rx - xx}|^{\frac{3}{2}}}$ by making $\dot{x}$ conftant. Then CI $= \frac{-\dot{x}\ddot{y}}{\dot{y}^3} \times Q$ is $= \frac{hr\dot{x}^3 Q}{\overline{2rx - xx}|^{\frac{3}{2}}} \times \frac{r^3}{h^3\dot{x}^3} \times \frac{\overline{2rx - xx}|^{\frac{3}{2}}}{\overline{r - x}|^3} = \frac{r^4 Q}{h^2 . \overline{r - x}|^3}$.
But when x is $= 0$, this expreffion becomes $a = \frac{rQ}{hh}$, and then $Q = \frac{ahh}{r}$; confequently CI is $= a \times \overline{\frac{r}{r - x}}|^3 = \frac{DR \times DQ^3}{PQ^3}$, the fame as in the circle.——And the fame expreffion may be brought out by making $\dot{y}$ conftant.

Hence the nature of the curve KI is thus expreffed, KR $= a + x - a \times \overline{\frac{r}{r - x}}|^3 = a + r - \frac{r}{h}\sqrt{hh - yy} - \frac{ah^3}{\overline{hh - yy}|^{\frac{3}{2}}}$, and is of the fame kind with that in the laft example.——But the elliptic arch may take a ftreight line at top better than the circular one, when the longer axe is hori-

horizontal, becauſe the arch is flatter, or of a leſs curvature; and worſe than the circular arch, when the ſhorter axe is horizontal.

EXAMPLE 3.

To determine the figure of the extrados of a parabolic arch of equilibration.

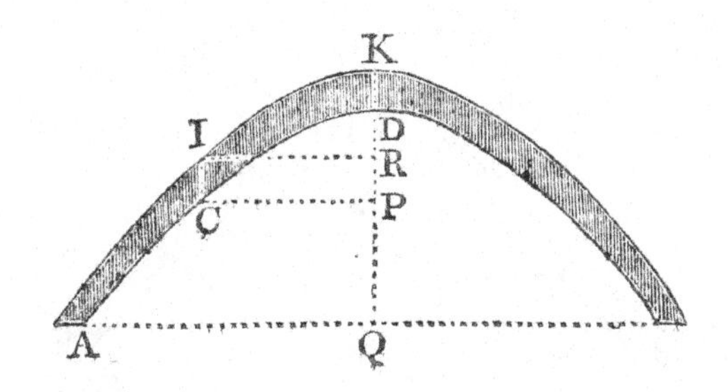

PUT a = KD, r = DQ, b = QA, x = DP, and y = PC = RI.

Then, by the nature of the curve, $bb : yy :: r : x = \frac{ryy}{bb}$; and hence $\dot{x} = \frac{2ry\dot{y}}{bb}$, and $\ddot{x} = \frac{2r\dot{y}}{bb}$, by making $\dot{y}$ conſtant. Then CI $= \frac{\ddot{x}}{\dot{y}^2} \times Q$ is $= \frac{2rQ}{bb}$ = a conſtant quantity = a; that is, CI is every-where equal to KD.

 Conſe-

Consequently KR is = DP; and since RI is = PC, it is evident that KI is the same parabolic curve with DC, and may be placed any height above it.

EXAMPLE 4.

To find the figure of the extrados for an hyperbolic arch of equilibration.

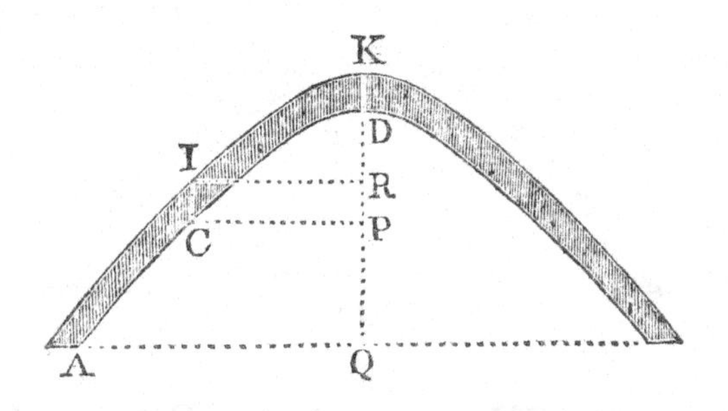

PUT a = KD, r = the semi-transverse, and b = the semi-conjugate axe, x = DP, and y = PC = RI.

Then, by the nature of the hyperbola, $y = \frac{b}{r}\sqrt{2rx + xx}$; hence $\dot{y} = \frac{b\dot{x}}{r} \times \frac{r + x}{\sqrt{2rx + xx}}$, and $\ddot{y} = \frac{-br\dot{x}^2}{\overline{2rx + xx}|^{\frac{3}{2}}}$, by making $\dot{x}$ constant.

Wherefore CI or $\frac{-\dot{x}\ddot{y}}{\dot{y}^3} \times Q = \frac{r^4 Q}{b^2 \times \overline{r+x}|^3}$. But

when

when $x = 0$, this expression becomes $\frac{rQ}{hh} = a$; hence $Q = \frac{ahh}{r}$, and consequently CI or $\frac{r^4 Q}{h^2 \times \overline{r+x}|^3}$ is $= \frac{ar^3}{\overline{r+x}|^3}$.

Whence the equation to the curve KI required will be KR $= (a + x - \text{CI} =)\ a + x - \frac{ar^3}{\overline{r+x}|^3} = a - r + \frac{r}{h}\sqrt{hh + yy} - \frac{ah^3}{\overline{hh+yy}|^{\frac{3}{2}}}$.

Scholium.

In this hyperbolic arch then, it is evident that the extrados KI continually approaches nearer to the intrados; whereas in the circular and elliptic arches, it goes off continually farther from it; and in the parabola, the two curves keep always at the same distance; observing however that by the distance between the two curves, in each of these cases, is meant their distance in the vertical direction.

Ex-

EXAMPLE 5.

To find the extrados for a catenarian arch of equilibration.

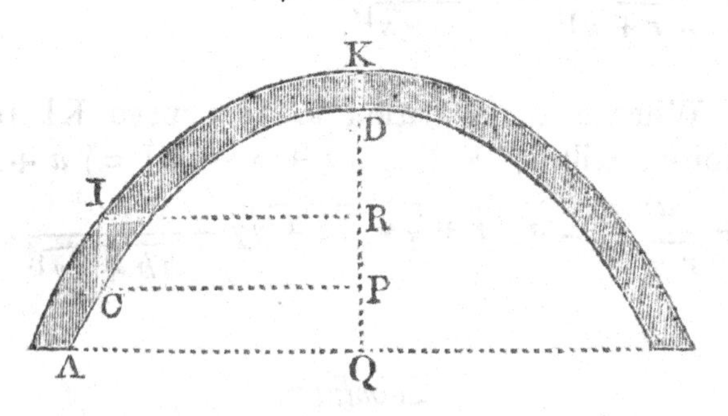

LET $a = KD$, $x = DP$, and $y = PC = RI$, as before; alſo let c denote the conſtant tenſion of the curve at the vertex.

Then, by the nature of the catenary, y is $= c \times$ hyp. log. of $\frac{c + x + \sqrt{2cx + xx}}{c}$; hence, taking the fluxions, we have $\dot{y} = \frac{c\dot{x}}{\sqrt{2cx + xx}}$, and $\ddot{y} = - c\dot{x}^2 \times \frac{c + x}{\overline{2cx + xx}|^{\frac{3}{2}}}$, by making $\dot{x}$ conſtant. Wherefore CI or $\frac{-\dot{x}\ddot{y}}{\dot{y}^3} \times Q$ is $= \frac{c + x}{cc} \times Q$. But at the vertex x is $= 0$, and CI $= a = \frac{Q}{c}$; conſequently Q is $= ac$. This being written

written for it, there results $CI = \frac{c + x}{c} \times a = a + \frac{ax}{c}$.

Hence, for the nature of the curve KI, we have $KR = (a + x - CI =)\ x - \frac{ax}{c} = \frac{c - a}{c} \times x$.

Corollary.

AND hence the abscissa DP is to the abscissa KR, always in the constant proportion of c to $c - a$. So that, when a is less than c, R and the curve KI lies below the horizontal line; but when a is greater than c, they lie above it; and when a is equal to c, KR is always equal to nothing, and KI or the extrados coincides with the horizontal line.

As a diminishes, the line KI approaches nearer to DC in all its parts, till when a entirely vanishes, or is so little in respect of c as to be omitted in the expression $\frac{c - a}{c} \times x = KR$, the two curves quite coincide throughout.

Scholium.

As we have found above that the extrados will be a streight horizontal line when a is equal to to c, I ſhall here make a calculation to determine, in that caſe, the value of c, and conſequently of a with reſpect to x and y, or a given ſpan and height of an arch.

Now the equation to the curve expreſſed in terms of c, x, and y, is $y = c \times$ hyp. log. of $\frac{c + x + \sqrt{2cx + xx}}{c}$; and when x and y are given, the value of c may be found from this equation, by the method of trial and error. But as the proceſs would be at beſt but a tedious one, and perhaps the method not eaſy in this caſe to be practiſed by every perſon, I ſhall here inveſtigate a ſeries for finding the value of c from thoſe of x and y in a direct manner.

Since then y is $= c \times$ hyp. log. of $\frac{c + x + \sqrt{2cx + xx}}{c}$, by taking the fluxion of this equation, we have $\dot{y} = \frac{c\dot{x}}{\sqrt{2cx + xx}} = \frac{\frac{1}{2}d\dot{x}}{\sqrt{dx + xx}}$ by writing d for $2c$; and by expanding this expreſſion into a ſeries, it becomes

$$\dot{y} = \tfrac{1}{2}\dot{x}\sqrt{\frac{d}{x}} \times$$

$$: 1 - \frac{x}{2d} + \frac{1.3x^2}{2.4d^2} - \frac{1.3.5x^3}{2.4.6d^3} + \frac{1.3.5.7x^4}{2.4.6.8d} \text{ \&c.}$$

and,

and, by taking the fluents we have $y = \sqrt{dx} \times$

$$: 1 - \frac{x}{2.3d} + \frac{1.3x^2}{2.4.5d^2} - \frac{1.3.5x^3}{2.4.6.7d^3} + \frac{1.3.5.7x^4}{2.4\ 6.8.9d^4} \text{ \&c.}$$

and hence, by dividing by x, we have $\frac{y}{x} = \sqrt{\frac{d}{x}} \times$

$$: 1 - \frac{x}{2.3d} + \frac{1.3x^2}{2.4.5d^2} - \frac{1.3.5x^3}{2.4.6.7d^3} + \frac{1.3.5.7x^4}{2.4.6.8.9d^4} \text{ \&c.}$$

or, by writing v for $\frac{y}{x}$ and w for $\sqrt{\frac{d}{x}}$, it is $v =$

$$w - \frac{1}{2.3w} + \frac{1.3}{2.4.5w^3} - \frac{1.3.5}{2.4.6.7w^5} + \frac{1.3.5.7}{2.4.6.8.9w^7} \text{ \&c.}$$

Then, by reverting this series, we have $w =$ $v + \frac{1}{6v} - \frac{37}{360v^3} + \frac{547}{5040v^5} - \frac{337}{5600v^7}$ &c. And hence, by squaring, &c. and restoring the original letters, it is $(\frac{1}{2}d = \frac{1}{2}xw^2 =)$ $c = \frac{1}{2}x \times$

$$: \frac{y^2}{x^2} + \frac{1}{3} - \frac{8x^2}{45y^2} + \frac{691x^4}{3780y^4} - \frac{22851x^6}{453600y^6} \text{ \&c.}$$

where a few of the first terms are sufficient to determine the value of c pretty nearly.

Now, for an example in numbers, suppose the height of the arch to be 40 feet, and its span 100, which are nearly the dimensions of the middle arch of Blackfriar's Bridge at London. Then $x = 40$, and $y = 50$; which being substituted for them in this series, it gives $c =$ 36·88 feet nearly. So that to have made that arch a catenarian one, with a streight line above, the top of the arch must have been almost of the immenſe thickneſs of 37 feet, to have kept it in equilibrium.

But if the height and ſpan be 40 and 100 feet, as above, and the thickneſs of the arch at top be aſſumed equal to 6 feet, then the extrados will not be a right line, but as it is drawn in the figure to this example, which figure is accurately conſtructed according to theſe dimenſions.

It may be farther remarked, that the curves in theſe laſt three examples, viz. the parabola, hyperbola, and catenary, are all very improper for the arches of a bridge conſiſting of ſeveral arches; becauſe it is evident from their figures, which are all accurately conſtructed, that all the building or filling up of the flanks of the arches will tend to deſtroy the equilibrium of them. But in a bridge of one ſingle arch whoſe extrados riſes pretty much from the ſpring to the top, one of theſe figures will anſwer better than any of the former ones.

Ex-

EXAMPLE 6.

To determine the extrados of a cycloidal arch of equilibration.

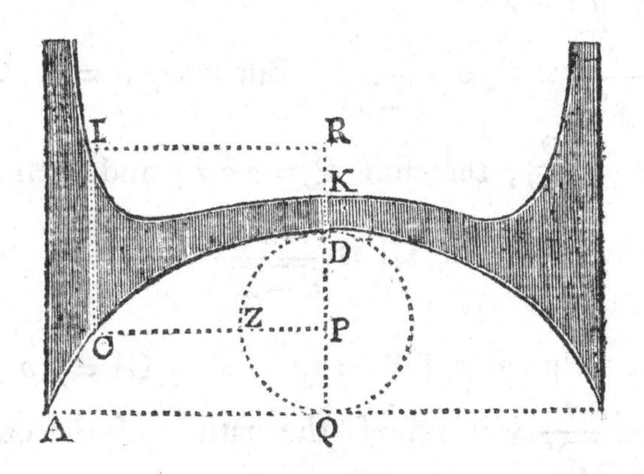

LET DZQ be the circle from which the cycloid DCA is generated, and the other lines as before.

Put $a =$ DK, $x =$ DP, and $y =$ PC $=$ RI; also put $d =$ DQ the diameter of the circle, and $z =$ the circular arc DZ.

Then, by the nature of the cycloid, CZ is always equal to DZ $= z$; and, by the nature of the circle, PZ is $= \sqrt{dx - xx}$; wherefore PC or $y =$ (CZ + ZP =) $z + \sqrt{dx - xx}$. Hence $\dot{y} = \dot{z} + \frac{\frac{1}{2}d - x}{\sqrt{dx - xx}} \times \dot{x}$; but $\dot{z} = \frac{\frac{1}{2}d\dot{x}}{\sqrt{dx - xx}}$

by

by the nature of the circle; therefore $\dot{y} = \frac{d - x}{\sqrt{dx - xx}} \times \dot{x} = \dot{x}\sqrt{\frac{d - x}{x}}$; and then $\ddot{y} = \frac{-d\dot{x}^2}{2x\sqrt{dx - xx}}$, making $\dot{x}$ constant. Hence CI $= \frac{-\dot{x}\ddot{y}}{\dot{y}^3} \times Q = \frac{\frac{1}{2}dQ}{\overline{d - x}|^2}$. But when $x = 0$, CI is $= a = \frac{Q}{2d}$; therefore $Q = 2ad$; and then the general value of CI is $\frac{add}{\overline{d - x}|^2}$.

Consequently KR $= (a + x - \text{CI} =)\ a + x - \frac{add}{\overline{d - x}|^2}$ will expreſs the nature of the curve KI; which reſembles that for the circle and ellipſe, as evidently appears by comparing the figures together, each of them being accurately conſtructed. But this figure ſeems to be rather better than either of them, as the extrados approaches rather nearer to a right line, and extends farther out before it is bent upwards.

Other examples of known curves might be given; but thoſe that have been put down already, ſeem to be the fitteſt for real practice; and there is a ſufficient variety among them, to ſuit the various circumſtances of convenience, ſtrength, and beauty.

I ſhall

I ſhall now proceed to another general problem, which is the reverſe of the laſt one, and determines the figure of the intrados for any given figure of the extrados, ſo that the arch may be in equilibrium in all its parts.

PROPOSITION V.

HAVING the Extrados given, to find the Intrados. That is, having given the nature or form of a line bounding the top of a wall above an arch; to find the figure of the arch, ſo that by the preſſure of the ſuperincumbent wall, the whole may remain in equilibrium.

Solution.

PUT a = DK the thickneſs of the arch at top, x = DP the abſciſſa of the intrados DC, z = KR the abſciſſa of the given extrados KI, and y = PC = RI their equal ordinates.

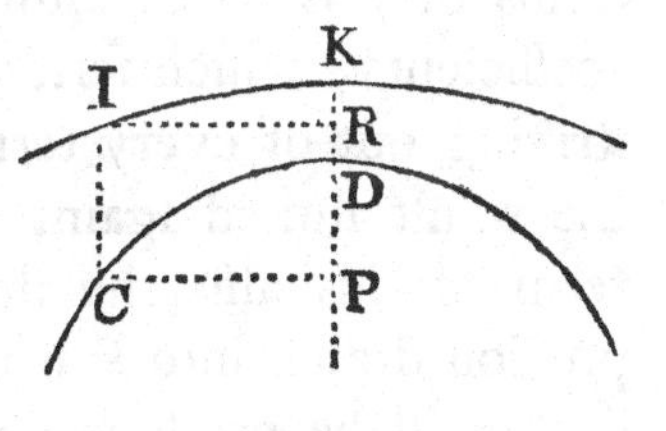

Then, by the laſt propoſition, CI is $= \frac{\dot{y}\ddot{x} - \dot{x}\ddot{y}}{\dot{y}^3} \times Q$; but CI is alſo evidently equal to $a + x - z$; therefore

therefore $a + x - z$ is $= \frac{\dot{y}\ddot{x} - \dot{x}\ddot{y}}{\dot{y}^3} \times Q = \frac{Q}{\dot{y}} \times$ the fluxion of $\frac{\dot{x}}{\dot{y}}$; where Q is a conſtant quantity, as uſed in the laſt propoſition, and always to be determined from the nature or conditions of each particular caſe.

Hence then, by ſubſtituting in this equation the given value of z inſtead of it, as expreſſed in terms of y, the reſulting equation will then involve only x and y together with their firſt and ſecond fluxions, beſides conſtant quantities. And from it the relation between x and y themſelves may be found, by the application of ſuch methods as may ſeem to be beſt adapted to the particular form of the given equation to the extrados. In general, a proper ſeries for the value of x in terms of y is to be aſſumed with indeterminate coefficients; which ſeries being put into fluxions, ſtriking out of every term the fluxion of y; and the reſult fluxed again, ſtriking out from every term of this alſo the fluxion of y; the laſt expreſſion drawn into Q being equated to $a + x - z$, there will be produced an equation from which will be found the values of the coefficients of the terms in the aſſumed value of x.

But in the particular caſe when z is always nothing, or the extrados a right horizontal line, a difa different and eaſier proceſs obtains, as in this following example.

EXAMPLE.

To find an arch of equilibration whoſe extrados ſhall be a horizontal line.

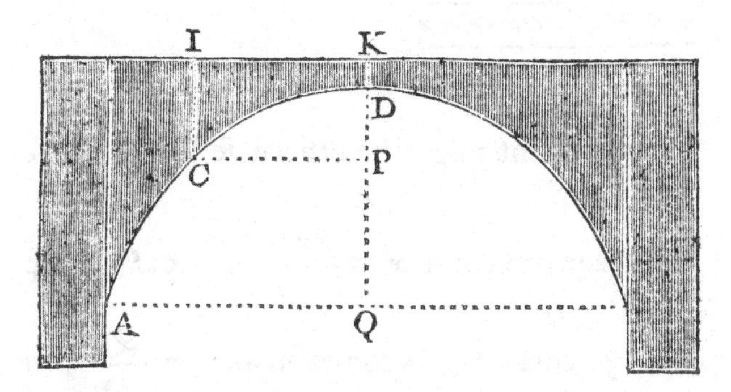

Making the notation as in the propoſition, we have $z = 0$, and therefore $a + x = \frac{Q}{\dot{y}} \times$ the fluxion of $\frac{\dot{x}}{\dot{y}}$.

Now aſſume $\dot{y} = \frac{\dot{x}}{v}$; then $\frac{\dot{x}}{\dot{y}} = v$, and $\frac{Q}{\dot{y}} \times$ flux. of $\frac{\dot{x}}{\dot{y}} = \frac{Qv\dot{v}}{\dot{x}}$; that is, $a + x = \frac{Qv\dot{v}}{\dot{x}}$; hence $a\dot{x} + x\dot{x} = Qv\dot{v}$. Then, by taking the fluents, we have $2ax + x^2 = Qv^2$; hence $v = \sqrt{\frac{2ax + xx}{Q}}$,

and

and consequently $\dot{y} = (\frac{\dot{x}}{v} =) \frac{Q^{\frac{1}{2}}\dot{x}}{\sqrt{2ax + xx}}$. Then the fluent of this is $y = Q^{\frac{1}{2}} \times$ hyp. log. of $2a + 2x + 2\sqrt{2ax + xx}$; but when $x = 0$, this is $Q^{\frac{1}{2}} \times$ hyp. log. of $2a$; therefore the correct fluent is $y = Q^{\frac{1}{2}} \times$ hyp. log. of $\frac{a + x + \sqrt{2ax + xx}}{a}$.

Or the fluent might be otherwise found thus.

THE equation $a + x = \frac{\dot{y}\ddot{x} - \dot{x}\ddot{y}}{\dot{y}^3} \times Q$, supposing $\dot{y}$ constant, becomes $a + x = \frac{Q\ddot{x}}{\dot{y}^2}$, or $a\dot{y}^2 + x\dot{y}^2 = Q\ddot{x}$; multiply by $\dot{x}$, and then $a\dot{x}\dot{y}^2 + x\dot{x}\dot{y}^2 = Q\dot{x}\ddot{x}$; and hence, by taking the fluents, $2ax\dot{y}^2 + x^2\dot{y}^2 = Q\dot{x}^2$; consequently $\dot{y}^2 = \frac{Q\dot{x}^2}{2ax + xx}$, or $\dot{y} = \frac{Q^{\frac{1}{2}}\dot{x}}{\sqrt{2ax + xx}}$. And then the rest will be as above.

Now the value of Q will be found by writing in this equation some particular correspondent known values of x and y: thus when P arrives at Q, then $x = DQ = r$, and $y = QA = h$; these being substituted for them, we have $h = Q^{\frac{1}{2}} \times$ hyp.

hyp. log. of $\frac{a + r + \sqrt{2ar + rr}}{a}$, and confe-

quently $Q^{\frac{1}{2}} = \frac{h}{\text{hyp. log. of } a + r + \frac{\sqrt{2ar + rr}}{a}}$.

Wherefore the general value of y is thus, $y =$

$$h \times \frac{\text{hyp. log. } \frac{a + x + \sqrt{2ax + xx}}{a}}{\text{hyp. log. } \frac{a + r + \sqrt{2ar + rr}}{a}}.$$

Hence, when $Q^{\frac{1}{2}}$ is $= a$, the curve DC is the catenary; and in general the ordinate is everywhere to the correfponding ordinate of the catenary whofe tenfion at the vertex is a, as h is to $a \times$ hyp. log. of $\frac{a + r + \sqrt{2ar + rr}}{a}$.

If x were defired in terms of y, it would be thus. Put $A =$ the hyp. log. of a, and $D = \frac{1}{h} \times$ hyp. log. of $\frac{a + r + \sqrt{2ar + rr}}{a}$; then $Dy + A =$ hyp. log. of $a + x + \sqrt{2ax + xx}$: Again, put $N =$ the number whofe hyp. log. is $Dy + A$; then $N = a + x + \sqrt{2ax + xx}$; and hence $x = \frac{\overline{N - a}|^2}{2N}$, or $a + x = \text{KP} = \frac{N^2 + a^2}{2N}$.

By taking $\text{AQ} = h = 50$, and $\text{DQ} = r = 40$, alfo $\text{DK} = a = 6$. Then the hyp. log.

G of

of $\frac{a + r + \sqrt{2ar + rr}}{a}$ is = the hyp. log. of $\frac{46 + 4\sqrt{130}}{6}$ = the hyp. log. of 15·26784 = 2·7257487; by which dividing $h = 50$, the quotient is 18·343584. So that the ordinate y will be conſtantly in that caſe equal to 18·343584 × the hyp. log. of $\frac{6 + x + \sqrt{12x + xx}}{6}$. Alſo $\frac{1}{18\cdot343584}$ = ·05451497 is = D, and A = hyp. log. of 6 = 1·7917594; then N = the number whoſe hyp. log. is 1·7917594 + ·05451497y. And then by aſſuming ſeveral values of one of the letters x, y, the correſponding values of the other will be found from one of the two equations above.

And in this manner were calculated the numbers in the following table; from which the curve being conſtructed, it will be as appears in the figure to the example.—— And thus we have an arch in equilibrium in all its parts, and its top a ſtreight line, as is generally required in moſt bridges; or at leaſt they are ſo near a horizontal line, that their difference from it will cauſe no ſenſible ill conſequence. It is alſo both both of a graceful figure, and of a convenient form for the paſſage through it. So that there can be no good reaſon for neglecting to uſe it in works of any conſequence.

The

The Table for Conſtructing the Curve in this Example.

Value of KI	Value of IC	Val. of KI	Value of IC	Val. of KI	Value of IC
0	6·000	21	10·381	36	21·774
2	6·035	22	10·858	37	22·948
4	6·144	23	11·368	38	24·190
6	6·324	24	11·911	39	25·505
8	6·580	25	12·489	40	26·894
10	6·914	26	13·106	41	28·364
12	7·330	27	13·761	42	29·919
13	7·571	28	14·457	43	31·563
14	7·834	29	15·196	44	33·299
15	8·120	30	15·980	45	35·135
16	8·430	31	16·811	46	37·075
17	8·766	32	17·693	47	39·126
18	9·168	33	18·627	48	41·293
19	9·517	34	19·617	49	43·581
20	9·934	35	20·665	50	46·000

The above numbers may be feet or any other lengths of which DQ is 40 and QA is 50. But when DQ is to QA in any other proportion than that of 4 to 5, or when DK is not to DQ as 6 to 40 or 3 to 20; then the above numbers will not anſwer; but others muſt be found by the ſame rule, to conſtruct the curve by.

In the beginning of the table, as far as 12, the value of KI is made to differ by 2, becauſe the value of IC in that part increafes fo very flowly.

Other examples of given extrados might be taken ; but as there can fcarcely ever be any real occafion for them, and as the trouble of calculation would be, in moft cafes, extremely great, they are omitted.

SEC

SECTION III.

Of the Piers.

PROPOSITION VI.

To find the diſtance QM *of the center of gravity of the given circular arc* AD, *from* DQ *the verſed ſine of the ſaid arc,* QA *being its right ſine.*

Solution.

PUT r = the radius, z = any arc DR, and x = its ſine TR or QS.

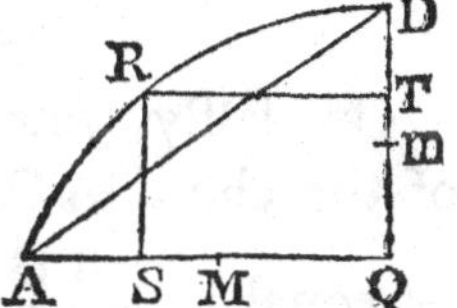

Then, by mechanics, the force of a particle $\dot{z}$ of the curve placed at R is TR $\times \dot{z} = x\dot{z}$; and the force of all the particles will be equal to the fluent of $x\dot{z}$; which muſt be equal to QM drawn into the whole line; that is, QM $\times z$ = the fluent of $x\dot{z}$, or QM $= \frac{1}{z} \times$ fluent of $x\dot{z}$. And this is a general theorem, whether z be a line, ſurface, or ſolid; ſuppoſing the two former to be affected with gravity.

Now,

Now, by the nature of the circle, $\dot{z} = \frac{r\dot{x}}{\sqrt{rr - xx}}$; and therefore $x\dot{z} = \frac{rx\dot{x}}{\sqrt{rr - xx}}$; the correct fluent of which is $r \times \overline{r - \sqrt{rr - xx}}$. Consequently QM is $= r \times \frac{r - \sqrt{rr - xx}}{z}$; which, when $x =$ QA, and $z =$ the arc AD, becomes $\text{QM} = r \times \frac{r - \sqrt{r^2 - \text{QA}^2}}{\text{ARD}} =$ the distance from DQ required.

Or, since $r - \sqrt{r^2 - \text{QA}^2}$ is $=$ QD, the same distance QM will be expressed by $r \times \frac{\text{DQ}}{\text{ARD}}$.

Or, lastly, since $r \times \text{QD}$ is half the square of the chord AD, the same distance QM will be equal to $\frac{\text{AD}^2}{2\text{ARD}}$ or $\frac{\text{AQ}^2 + \text{QD}^2}{2\text{ARD}}$.

Corollary.

When ARD is a quadrant, then $\text{AQ} = \text{QD} = r$, and the rule is $\text{QM} = (\frac{rr}{\text{ARD}} = \frac{rr}{\cdot 7854 \times 2r} =) \frac{r}{1\cdot 5708}$. Or $\text{QM} = \frac{2}{3}r$ nearly, or $= \frac{7}{11}r$ extremely near.

PROPO-

PROPOSITION VII.

THE figure being the same as in the last proposition, in is required to find the distance Qm *of the center of gravity of the arc* ARD *from the sine* AQ.

Solution.

As in the last proposition, QM will be $= \frac{1}{\text{AR}}$ $\times$ the fluent of SR $\times$ $\dot{\text{A}}$R.

But, putting $z = \text{AR}$, $x = \text{QS} = \text{RT}$, $r =$ the radius, $b = \text{DQ}$, and $s = \text{QA}$, we shall have $\dot{z} = \dot{\text{A}}\text{R} = \frac{-r\dot{x}}{\sqrt{rr - xx}}$, and $\text{SR} = b - r + \sqrt{rr - xx}$; hence $\text{SR} \times \dot{\text{A}}\text{R} = \frac{r - b}{\sqrt{rr - xx}} \times r\dot{x} - r\dot{x} = \overline{b - r} . \dot{z} - r\dot{x}$; the correct fluent of which is $\overline{b - r} . z + \overline{s - x} . r$.

Consequently Qm is $= b - r + \frac{s - x}{z} . r = b - r + \frac{\text{AS}}{z} . r$ And when R arrives at D, it is $\text{Qm} = b - r + \frac{sr}{A}$.

Or,

Or, ſince r is $= \frac{ss + hh}{2h}$, the ſame diſtance Qm will be $= \frac{hh - ss}{2h} + \frac{hh + ss}{2h} \cdot \frac{s}{A}$; where A is the whole arc ARD.

Corollary.

When ARD is a quadrant, then h and s are each $= s$, and the rule is $\frac{rr}{A}$, the ſame as in the corollary to the laſt.

PROPO-

PROPOSITION VIII.

To find the distance QM *of the center of gravity of the space* AIKDSA *from* KQ; *supposing* DA *to be a circular arc whose sine is* AQ, *its versed sine* QD, *and* AI, IK, *parallel to* DQ, QA *respectively.*

Solution.

Draw RS, ST parallel to DQ, QA. And put $a =$ DK, $r =$ VD = VW the radius of the circle, $x =$ TS = KR, and $z =$ the area DSRK.

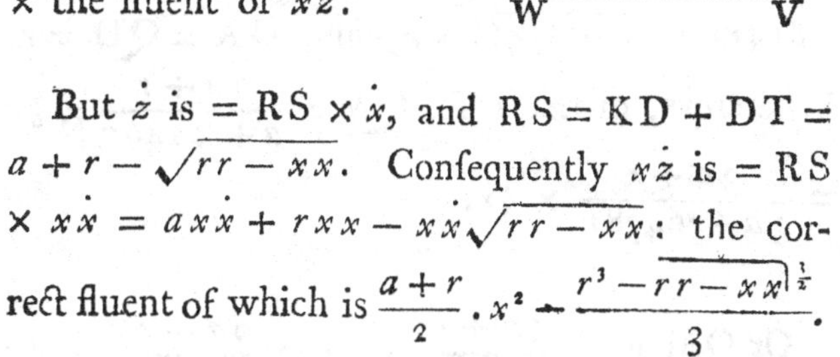

Then, as in prop. 6, we shall have QM $= \frac{1}{z}$ $\times$ the fluent of $x\dot{z}$.

But $\dot{z}$ is = RS $\times \dot{x}$, and RS = KD + DT = $a + r - \sqrt{rr - xx}$. Consequently $x\dot{z}$ is = RS $\times x\dot{x} = ax\dot{x} + rx\dot{x} - x\dot{x}\sqrt{rr - xx}$: the correct fluent of which is $\frac{a+r}{2} . x^2 - \frac{r^3 - \overline{rr - xx}|^{\frac{3}{2}}}{3}$.

Wherefore QM is $= \frac{a+r}{2z}.x^2 - \frac{r^3-\overline{\sqrt{rr-xx}}|^3}{3z}$

$= (\frac{a+r}{2}.\frac{xx}{z} - \frac{rxx+\overline{r-\sqrt{rr-xx}}.\overline{rr-xx}}{3z}$

$= \frac{3a+r}{6}.\frac{xx}{z} - \frac{TD}{3}.\frac{\overline{r-TD}|^2}{z} =)$

$\frac{\overline{3a+r}.TS^2 - 2TD.\overline{r-TD}|^2}{6z}$ or $= \frac{r^2-y^2}{2z}.m$

$- \frac{r^3-y^3}{3z}$, putting m = VK and y = VT. And when SR arrives at AI, then QM is

$= \frac{\overline{3a+r}.QA^2-2QD.\overline{r-QD}|^2}{6AIKDSA} = \frac{r^2-VQ^2}{2A}.VK$

$- \frac{r^3-VQ^3}{3A}$; putting A for the whole ſpace AIKDW.

Corollary 1.

WHEN DA is a quadrant; then the ſpace AIKDSA or AIKQ − ASDQ is $= \overline{a+r}.r - \cdot7854rr = \overline{a-\cdot2146r}\times r$, and QA = QD = r. Wherefore, in that caſe, QM $= \frac{3a+r}{a+\cdot2146r}\times\frac{1}{6}r$

$= \frac{3a+r}{3a+\cdot6438r}\times\frac{1}{2}r$.

Or QM is $= \frac{3a+r}{3a+\frac{2}{3}r}\times\frac{1}{2}r = \frac{9a+3r}{9a+2r}\times\frac{1}{2}r$ nearly.

nearly. Or, rather, it is $= \frac{3a + r}{3a + \frac{9}{14}r} \times \frac{1}{2}r =$ $\frac{42a + 14r}{42a + 9r} \times \frac{1}{2}r$ extremely near.

Corollary 2.

When a is nothing, then (AD being a quadrant) QM is $= \frac{r}{1 \cdot 2876}$. Or it is $\frac{7}{9}r$ very nearly.

And when a is $= \frac{1}{9}r$, then QM is $= \frac{r}{1 \cdot 4657}$. Or $\frac{15}{22}r$ very nearly.

Lastly, when $a = \frac{1}{15}r$, which is nearly the proportion in pretty large arches; then QM is $= \frac{r}{1 \cdot 406}$. Or $\frac{5}{7}r$ very nearly.

Proposition IX.

To find the distance of the center of gravity of the space kiDSA *from the sine* QA *of the circular arc* ASD; *where* ki *is perpendicular to* QAk, *and the rest of the lines as in the last figure.*

Solution.

Put $a =$ kA, $s =$ AQ, $m =$ kQ $= a + s$, $r =$ VW $=$ VD the radius, $z =$ any variable space krSA, and $x =$ TS the sine of the arc SD. Also $A =$ the space kiDSA.

 Then

Then rS $= m - x$, and, by the circle, VT $= \sqrt{rr - xx}$; hence $\dot{z} = \text{rS} \times \dot{\text{V}}\text{T} = \frac{m - x}{\sqrt{rr - xx}} \times -x\dot{x}$; consequently $\text{VT} \times \dot{z} = \overline{m - x} \times -x\dot{x}$; the correct fluent of which is $\frac{s^2 - x^2}{2} \times m - \frac{s^3 - x^3}{3}$. Wherefore the distance from VW is $\text{V}m = \frac{s^2 - x^2}{2z} \times m - \frac{s^3 - x^3}{3z}$ for the general space krSA.

And when S arrives at D, x is $= 0$; and then Vm is $= \frac{s^2 m}{2A} - \frac{s^3}{3A} = \frac{3m - 2s}{6A} \times s^2 = \frac{3a + s}{6A} \times s^2 = \frac{3\text{kA} + \text{AQ}}{6\text{kiDSA}} \times \text{AQ}^2 =$ the distance of the center of gravity from VW.

Corollary.

WHEN *A* coincides with *W*, or the arc a quadrant, then s is $= r$; and the rule becomes as in Corollary 1 to the last. Also the 2d Corollary to that may be understood here, making the same suppositions as in it.

Scholium.

THE four preceding propositions are premised as necessary to the examples to the following general one, which determines the thickness of the

the piers neceſſary to reſiſt the ſpread or ſhoot of any given arch, and that whether the whole or part or none of it is immerſed in water. Inſtances only of circular arcs are here given; becauſe that in determining the drift of the arch, whatever its curve may be, it will make little or no difference by ſuppoſing it to be circular.

PROPOSITION X.

To find the thickneſs of the piers of an arch, neceſſary to keep the arch in equilibrium, or to reſiſt its ſhoot or drift; independent of any other arches.

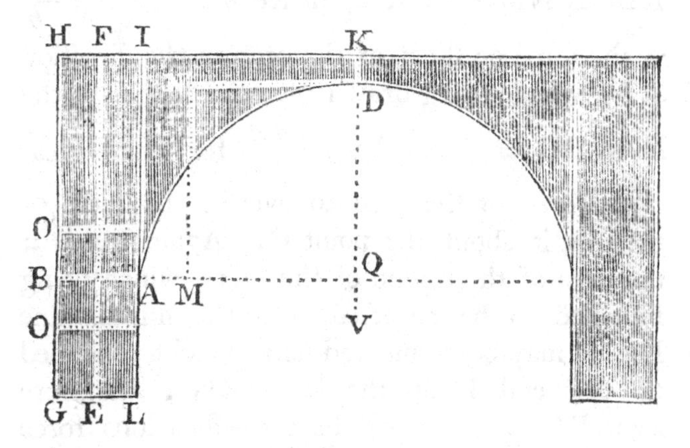

Solution.

LET IKDA be the half arch, and IHGL the pier to ſupport it, moveable about the point G, and biſected by the perpendicular EF.

Through

Through the center of gravity of the arch AIKD draw MN perpendicular to AQ the ſemiſpan, and meeting DN drawn parallel to AQ in N. And continue QA to meet GH in B.

Put a = DK, b = DQ = MN, c = AM, A = the area or ſection AIKD of the arch, d = AL = BG, e = FE, and x = AB = GL the required breadth of the pier.

Now (by prop. 63 *Emer.* Mechan.) the weight of the arch is to its preſſure in the direction AB, as NM is to MA; hence $b : c :: A : \frac{cA}{b}$ = the force or ſhoot of the arch in the direction AB; which being drawn into the length of the lever BG = d, we have $\frac{cdA}{b}$ for the efficacious force of the arch to overſet the pier, or to turn it about the point G. Again, ex is = the area of the ſection of the pier; which being ſuppoſed to be collected into the middle line EF, it may be conſidered as a weight appended to the end E of the lever EG; therefore $ex \times EG = \frac{1}{2}exx$ will be the efficacious force of the pier to prevent its being overturned. And that the arch and pier may be juſt kept in equilibrium, we muſt make the force and reſiſtance equal to each other, that is $\frac{1}{2}exx$

=

$= \frac{cdA}{h}$. Hence then $x = \sqrt{\frac{2cdA}{eh}} = \sqrt{\frac{2\,\mathrm{AM} \times \mathrm{AL} \times A}{\mathrm{DQ} \times \mathrm{EF}}}$ will be the breadth or thickness of the pier required.

In the above investigation it is supposed that the whole of the pier was out of water: But if any part of it OL be supposed to be immersed in water, that part will lose so much of its weight as is equal to its bulk of water; and since the specific gravity of water is to that of common stone, as 1 is to $2\frac{1}{2}$, or as 2 to 5, it is evident that OL will lose 2 parts in 5 of its weight. Hence then, putting $g = \mathrm{OG}$, since $\mathrm{OG} \times \mathrm{GL} = gx$ is the area immersed, therefore $\frac{2}{5}gx =$ the weight lost by the immersion; which being taken from ex the whole, we shall have $ex - \frac{2}{5}gx$ as the weight remaining appended to E; then this being drawn into GE $= \frac{1}{2}x$, and the product equated to the efficacious force of the arch as before, we have $\frac{1}{2}exx - \frac{1}{5}gxx = \frac{cdA}{h}$; and hence $x = \sqrt{\frac{10cdA}{h \cdot \overline{5e - 2g}}}$ for the thickness of the pier when it is immersed in water to the height expressed by g. ——Or, because g will be nearly equal to d, the theorem for the thickness may be $x = \sqrt{\frac{10cdA}{h \cdot \overline{5e - 2d}}} = \sqrt{\frac{10\,\mathrm{AM} \times \mathrm{AL} \times A}{\mathrm{DQ} \times \overline{5\mathrm{EF} - 2\mathrm{AL}}}}$.

Corol.

Corollary 1.

When DA is a quadrant, the arch is a complete ſemicircle; and then h is $= r$, $A = \overline{a + \frac{3}{14}r} \times r$ as in Cor. 1 to prop. 8, and by the ſame Corollary c or $r -$ QM is $= r - \frac{3a + r}{a + \frac{3}{14}r} \times \frac{1}{6}r$ $= \frac{3a + \frac{2}{7}r}{a + \frac{3}{14}r} \times \frac{1}{6}r$. Conſequently cA is $= \overline{3a + \frac{2}{7}r} \times \frac{1}{6}rr$.

This value being ſubſtituted in the two preceding theorems, we have $x = \sqrt{\frac{dr}{e}} \times \frac{21a + 2r}{21}$ $= \sqrt{\frac{dr}{a + r + d} \times \frac{21a + 2r}{21}} =$ $\sqrt{\frac{AL \times AQ}{IL} \times \frac{21 DK + 2 AQ}{21}}$ = thickneſs of the pier when it it is dry.——Or, if n expreſs what part a is of r, or DK = $\frac{1}{n}$th of DQ or QA, the ſame thickneſs will be $r\sqrt{\frac{d}{21} \times \frac{2 + 2n}{r + r + d.n}}$ $= AQ \times \sqrt{\frac{1}{21}AL \times \frac{21 + 2n}{AQ + \overline{AQ + AL}.n}}$.

And the thickneſs when AL is under water will be $x = \sqrt{\frac{5dr}{5e - 2d} \times \frac{21a + 2r}{21}} =$

$\sqrt{}$

$\sqrt{\frac{dr}{a+r+\frac{1}{5}d} \times \frac{21a+2r}{21}} =$
$\sqrt{\frac{AL \times AQ}{IL - \frac{2}{5}AL} \times \frac{21DK+2AQ}{21}}$.—Or, if $a = \frac{r}{n}$ as before, the same thickness will be
$r\sqrt{\frac{d}{21} \times \frac{21+2n}{r+r+\frac{1}{5}d.n}} = AQ \times$
$\sqrt{\frac{1}{21}AL \times \frac{21+2n}{AQ+AQ+\frac{1}{5}AL.n}}$.

Corollary 2.

When HG is = BG in the last figure; then the arch and pier will be as in this annexed figure. And, e being then $= d$, the two general theorems will become $x = \sqrt{\frac{2cA}{b}} = \sqrt{\frac{2A \times AM}{DQ}}$ for the thickness of the pier when dry, and $x = \sqrt{\frac{10cA}{3b}} = \sqrt{\frac{10A \times AM}{3DQ}} =$ the thickness when under water.

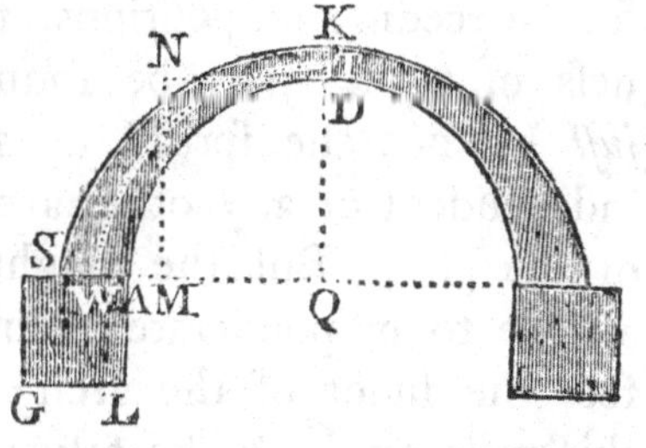

So that, in this case, it makes no difference of whatever height LA the pier is to the springing

 of

of the arch. For though the drift of the arch be increafed with the length of the lever or height of the pier, the weight of the pier itfelf, which acts againft it, is alfo increafed in the fame proportion.

Scholium.

IN the inveftigation of this propofition, the fections of the arch and pier are ufed for their folidities, as being evidently in the fame proportion, or in that of their weights, fince they are of the fame length, viz. the breadth of the bridge.

By the above rules, together with thofe in the four preceding propofitions, the neceffary thicknefs of a pier may be found, fo that it fhall *juft* balance the fpread or fhoot of the arch, independent of any other arch on the other fide of the pier. But the weight of the pier ought a little to preponderate againft or exceed in effect the fhoot of the arch; and therefore the thicknefs ought to be taken a little more than what will be found by thefe rules; unlefs it be fuppofed that the pointed projections of the piers againft the ftream, beyond the common breadth of the bridge, will be a fufficient addition to the pier, to give it the neceffary preponderancy.——But there is one very material thing, on account of which the thicknefs of the piers may be much diminifhed; viz. by the ftones ſtones of the wall above the vouſſoirs being bonded in with thoſe of the pier and with one another, the pier will carry part of their weight; which will not only diminiſh the weight of the whole arch and wall, but will alſo both add the ſame to the weight of the pier, and lengthen the lever E G, by moving the center of gravity a little nearer to L; but then alſo M will be a little nearer to Q, ſo that A M will be longer, and the effects of the change of the centers of gravity may be ſuppoſed nearly to balance each other.——In the foregoing propoſitions I have conſidered circular arches only, as it will make no difference of any conſequence, to ſuppoſe the arches of any other curve of the ſame ſpan and pitch. But this 10th prop. is general for all curves.

I ſhall now add a few examples of the calculation in numbers, to ſhew the manner, and in them alſo to point out the eaſieſt methods of calculation.

EXAMPLE I.

SUPPOSING the arch in the figure to the propoſition to be a ſemicircle whoſe height or pitch is 45 feet, and conſequently its ſpan 90 feet; alſo ſuppoſe the thickneſs D K at top to be 6 feet, and the height L A to the ſpringing 18; and let it be required to find the thickneſs G L of the pier neceſſary to reſiſt the drift of the arch.

This will be immediately found by Cor. 1, in which AQ is = 45, AL = 18, and $n = \frac{r}{a} = \frac{45}{6} = 7\frac{1}{2}$.

Then the firſt expreſſion $AQ \times \sqrt{\frac{AL}{21} \times \frac{21 + 2n}{AQ + AQ + AL.n}}$ will become $\frac{540}{\sqrt{2415}}$ = 10·988, or 11 feet nearly for the thickneſs of the pier when dry.

And the latter expreſſion $AQ \times \sqrt{\frac{AL}{21} \times \frac{21 + 2n}{AQ + AQ + \frac{3}{5} AL.n}}$ will give $\frac{540}{\sqrt{2163}}$ = 11·61 feet for the thickneſs when 18 feet are under water.

EXAMPLE 2.

In the ſame figure, ſuppoſe the ſpan to be 100 feet, the height 40 feet; alſo the thickneſs at top 6 feet, and the height of the pier to the ſpringer 18 feet as before.

Here the figure either is or may be conſidered as a ſcheme arch, or the ſegment of a circle, in which the verſed ſine QD is = 40, and the right ſine QA = 50; alſo DK = 6, AL = 18, EF = 64.

Now

Now, by the nature of the circle, the radius $VD = r$ is $= \frac{QA^2 + QD^2}{2QD} = \frac{50^2 + 40^2}{80} = 51\frac{1}{4}$; hence $VQ = 51\frac{1}{4} - 40 = 11\frac{1}{4}$; and the area of the semi-segment ADQ will be found to be 1490·9998, or 1491 nearly; which being taken from the rectangle $AIKQ = AQ \times QK = 50 \times 46 = 2300$, there remains $809 = A$ the area AIKD. Then, by prop. 8, QM will be $= \frac{VD^2 - VQ^2}{2A} \times VK - \frac{VD^3 - VQ^3}{3A} = \frac{51{\cdot}25^2 - 11{\cdot}25^2}{2 \times 809} \times 57\frac{1}{4} - \frac{51{\cdot}25^3 - 11{\cdot}25^3}{3 \times 809} =$ 33·58; and consequently $MA = AQ - QM = 50 - 33{\cdot}58 = 16{\cdot}42$.

Then, the first expression $\sqrt{\frac{2AL \times AM \times A}{DQ \times EF}}$ will become $\sqrt{\frac{36 \times 16{\cdot}42 \times 809}{40 \times 64}} = 13{\cdot}67$, or $13\frac{2}{3}$ feet nearly = the thickness of the pier when dry.

And the latter expression $\sqrt{\frac{10AL \times AM \times A}{DQ \times \overline{5EF - 2AL}}}$ will give $\sqrt{\frac{180 \times 16{\cdot}42 \times 809}{40 \times \overline{320 - 36}}} = 14{\cdot}508$, or $14\frac{1}{2}$ feet nearly = the thickness when 18 feet are under water.

EXAMPLE 3.

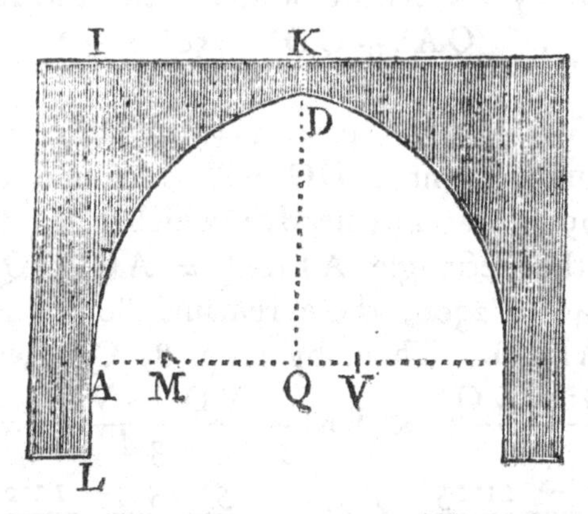

LET the arch be of the gothic kind, as in the annexed figure; in which DA is a circular arc whoſe center is V, its ſine DQ = 50 = the height of the arch, its verſed ſine AQ = 40 = the ſemi-ſpan, the thickneſs at top DK = 6, and the height AL of the pier to the ſpring = 18 as before.

Here the radius VA = $51\frac{1}{4}$ as in the laſt example, and the ſemi-ſegment ADQ = 1491, alſo the ſame as in the laſt example; then the rectangle IQ is = AQ × QK = 40 × 56 = 2240; from which taking the ſemi-ſegment, there remains 749 = *A* for the area AIKD. Then, by prop. 9, VM will be equal to $\frac{3KD + DQ}{6A} \times DQ^2 = \frac{18 + 50}{6 \times 749} \times 50^2 = 37{\cdot}83$; and hence MA $= c = 51{\cdot}25 - 37{\cdot}83 = 13{\cdot}42$.

Then

Then the first expression $\sqrt{\frac{2AL \times AM \times A}{DQ \times IL}}$ will become $\sqrt{\frac{36 \times 13{\cdot}42 \times 749}{50 \times 74}} = 9{\cdot}889$, or nearly 10 feet for the thickness of the pier when when it is all out of water.

And the latter one $\sqrt{\frac{10AL \times MA \times A}{DQ \times \overline{5IL - 2AL}}}$ will give $\sqrt{\frac{180 \times 13{\cdot}42 \times 809}{50 \times \overline{370 - 36}}} = 10{\cdot}409$, or $10\frac{1}{2}$ nearly = the thickness when 18 feet are under water.

EXAMPLE 4.

WHEN the arch stones only are laid, and the pier built no higher than the spring, it will appear as in the figure to corollary 2. And then if, in the first case, the arch be a complete semicircle whose diameter is 90 feet, and the thickness everywhere DK = AS = 6 feet: It is required to find the breadth of the piers.

The bounding arcs being quadrants, the area ADKS will be $\frac{AD + KS}{2} \times DK = \frac{90 + 102}{2} \times \frac{11}{14} \times 6 = 144 \times \frac{22}{7} = 452{\cdot}4 = A$. Now if TW be another concentric quadrant bisecting the area ADKS, the center of gravity of TW may

may be taken for that of the ſaid area. And then, by the corollary to prop. 6, QM will be $= \frac{7}{11}$QT; but ſince the quadrants QDA, QTW, QKS are in arithmetic progreſſion, the ſquares of their ſemidiameters QD, QT, QK will be in the ſame progreſſion, that is $2QT^2 = QD^2 + QK^2$, or $QT = \sqrt{\frac{QD^2 + QK^2}{2}} = \sqrt{\frac{45^2 + 51^2}{2}}$ $= 48{\cdot}094$; hence then $QM = \frac{7}{11}QT$ is $= \frac{7}{11} \times 48{\cdot}094 = 30{\cdot}605$, and conſequently $MA = 45 - 30{\cdot}6 = 14{\cdot}4$.

Then the former of the two expreſſions in corollary 2 to this propoſition, will give GL or $\sqrt{\frac{2A \times AM}{DQ}} = \sqrt{\frac{904{\cdot}8 \times 14{\cdot}4}{45}} = 17{\cdot}016$, or 17 feet for the thickneſs of the pier when out of water.

And the latter one $\sqrt{\frac{10A \times AM}{3DQ}}$ will become $\sqrt{\frac{4524 \times 14{\cdot}4}{135}} = 21{\cdot}97$, or nearly 22 feet for the thickneſs when the pier is immerſed in water.

Scholium.

OR, becauſe QT is nearly an arithmetic mean between QD and QK, half the ſum of QD and QK might have been uſed inſtead of it, without

without causing any sensible difference in the conclusion.

We might also exhibit general theorems for the thickness, in terms of the radius only. For, taking QT or QW $= \frac{QD + QK}{2}$, by the corollary to prop. 6 we have QM $= \frac{7}{11}$QW $= \frac{QD + QK}{22} \times 7$, and thence AM $= c =$ AQ $-$ QM $=$ QA $- \frac{7QD + 7QK}{22} = \frac{8QD - 7DK}{22}$.

Also $A = \frac{AD + KS}{2} \times DK = \frac{QD + QK}{2} \times \frac{11}{7} DK$ $= \frac{2QD + DK}{14} \times 11 DK$. Then these values being substituted in the expression $\sqrt{\frac{2A \times AM}{QD}}$ we shall have $\sqrt{\frac{16QD^2 - 6QD \times DK - 7DK^2}{14QD}}$ $\times$ DK for the thickness of the pier when dry; and the same expression multiplied by $\sqrt{\frac{5}{3}}$ will give the thickness when the pier is immersed in water. And, farther, if DK be assumed equal to any part of DQ, as DK $= n \times$ DQ; then the thickness in the former case will be QD $\times$ $\sqrt{\frac{16 - 6n - 7nn}{14}} \times n$, and in the latter QD $\times$ $\sqrt{\frac{80 - 30n - 35nn}{42}} \times n$.

K Then

Then, by aſſuming ſeveral values of *n* from $\frac{1}{10}$ to $\frac{1}{5}$, which are beyond the limits of it, the ſeveral breadths of the piers correſponding to the ſeveral values of the thickneſs of the arch, both when the pier is ſuppoſed to be out of water, and immerſed in it, will be found from theſe expreſſions as in the following table; where the fractional part $\frac{1}{7\frac{1}{2}}$ or $\frac{2}{15}$ is alſo given, becauſe it is the moſt common proportion.

A Table of the Breadth or Thickneſs of a Pier anſwering to the ſeveral thickneſſes of a ſemicircular arch, as in the foregoing example, QD *being the radius or ſemi-ſpan.*

For the pier dry		For the pier in water	
Thickneſs of the arch	Thickneſs of the pier	Thickneſs of the arch	Thickneſs of the pier
$\frac{1}{10}$QD	·331 QD	$\frac{1}{10}$QD	·427 QD
$\frac{1}{9}$QD	·348 QD	$\frac{1}{9}$QD	·449 QD
$\frac{1}{8}$QD	·368 QD	$\frac{1}{8}$QD	·475 QD
$\frac{1}{7\frac{1}{2}}$QD	·379 QD	$\frac{1}{7\frac{1}{2}}$QD	·488 QD
$\frac{1}{7}$QD	·391 QD	$\frac{1}{7}$QD	·505 QD
$\frac{1}{6}$QD	·420 QD	$\frac{1}{6}$QD	·542 QD
$\frac{1}{5}$QD	·455 QD	$\frac{1}{5}$QD	·588 QD

EXAMPLE 5.

BUT supposing the same figure in Cor. 2 to be a circular segment, whose chord or span is 100 feet, and height 40 feet, also the thickness of the arch 6 feet: To find the thickness of the piers.

Here the radius of the middle arc TW is $\frac{QW^2 + QT^2}{2QT} = \frac{53^2 + 43^2}{86} = 54\frac{7}{43}$; hence TW is an arc of 78° 6′, and its length will be 73·8293; which being multiplied by DK = AS = 6, we have $A = 442·9758$. Then, by prop. 6, QM will be found $= \frac{53^2 + 43^2}{2 \times 73·8293} = 31·545$. And consequently AM = 50 − 31·545 = 18·455.

Hence, by Cor. 2, it will be $\sqrt{\frac{2A \times AM}{DQ}} = \sqrt{\frac{885·9516 \times 18·455}{40}} = 20·218$ = the thickness of the pier when dry.

And $\sqrt{\frac{10A \times AM}{3DQ}} = \sqrt{\frac{4429·758 \times 18·455}{120}}$ = 26·101 = the thickness in water.

Otherwise.

BUT if the arch be suppofed to increafe in thicknefs from the top at D, where it is 6 feet, all the way to the fpring, where it is AS = 12 feet fuppofe; the height and fpan being 40 and 100 as before.

Then QS = 62, QW = 56, and QT = 43. Hence the radius of the arc TW will be $\frac{QW^2 + QT^2}{2QT} = \frac{56^2 + 43^2}{86} = 59{\cdot}3452$; and therefore TW is an arc of 70° 40, and its length = 73·1945. Confequently the area ADKS or $TW \times \frac{DK + AS}{2}$ will be $73{\cdot}1945 \times 9 = 658{\cdot}75 = A$. And, by prop. 6, QM will be $\frac{56^2 + 43^2}{2 \times 73{\cdot}1945} = 34{\cdot}053$; and therefore $AM = 50 - 34{\cdot}053 = 15{\cdot}947$.

Hence, as above, $\sqrt{\frac{1317{\cdot}5 \times 15{\cdot}947}{40}} = 22{\cdot}918$ will be the thicknefs of the pier when dry.

And $\sqrt{\frac{6587{\cdot}5 \times 15{\cdot}947}{120}} = 29{\cdot}588$ = the thicknefs in water.

Ex.

EXAMPLE 6.

In a gothic arch whose thickness at top is 6, the span 80, and height 50 feet; to find the thickness of the piers.

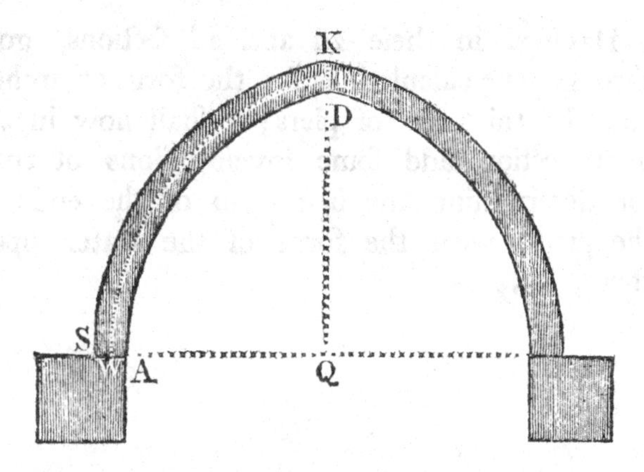

By the last example, TW is = 73·8293, its radius = $54\frac{7}{43}$, and the area ADKS = 442·9758. Then, by prop. 7, we have QM = 43 − $54\frac{7}{43}$ + $\frac{53 \times 54\frac{7}{43}}{73{\cdot}8293}$ = 27·718; and hence AM = 40 − 27·718 = 12·282.

Then, by Cor. 2, we shall have $\sqrt{\frac{2 \times 442{\cdot}9758 \times 12{\cdot}282}{50}}$ = 14·752 for the thickness of the pier when dry.

And $\sqrt{\frac{4429{\cdot}758 \times 12{\cdot}282}{150}}$ = 19·045 = the thickness when in water.

Also

Alſo if the arch ſtones were ſuppoſed to lengthen all the way from the top towards the lower end, the calculation might be made as in the laſt example.

Having, in theſe 2d and 3d ſections, gone through the calculations for the form of arches, and the thickneſs of piers; I ſhall now in the next ſection add ſome inveſtigations of rules for determining the beſt form of the ends of the piers, with the force of the water upon them, &c.

SECTION IV.

The Force of the Water, &c.

PROPOSITION XI.

To determine the form of the ends of a pier, so as to make the least resistance to, or be the least subject to the force of the stream of water.

Solution.

LET the following figure represent a horizontal section of the pier, AB its breadth, CD the given length or projection of the end, and ADB the line required, whether right or curved; also let EF represent the force of a particle of water acting on AD at F in the direction parallel to the axe CD; produce EF to meet AB in G, and draw the tangent FH, also draw EH perpendicular to FH, HI perpendicular to EF, and FK perpendicular to DC.

Now

Now the abſolute force EF of the particle of water may be reſolved into the two forces EH, HF, and in thoſe directions; of theſe the latter one, acting parallel to the curve, is of no effect; and the former EH is reſolved into the two EI, IH; ſo that EI is the efficacious force of the particle to move the pier in the direction of its axe or length: That is, the abſolute force is to the efficacious force, as EF is to EI.—— Then, ſince EF is the diameter of a ſemicircle paſſing through H, by the nature of the circle we ſhall have EF : EI :: $EF^2 : EH^2$ :: (by ſimilar triangles) $HF^2 : HI^2$ and :: the ſquare of the fluxion of the curve or line : the ſquare of the fluxion of the ordinate FK, becauſe HF, HI are parallel to the line and ordinate.

Wherefore, putting the abſciſſa DK $= x$, the ordinate KF $= y$, and the line DF $= z$, we ſhall have as $\dot{z}^2 : \dot{y}^2 :: 1$ (the force EF : $\frac{\dot{y}^2}{\dot{z}^2}$) $=$ the force of the particle at F to move the pier in the direction EFG. But the number of particles ſtriking againſt the indefinitely ſmall part of the line, is as $\dot{y}$; this drawn into the above found force of each, we have $\frac{\dot{y}^3}{\dot{z}^2} = \frac{\dot{y}^3}{\dot{x}^2 + \dot{y}^2}$ for the fluxion of the force, or the force acting againſt the part $\dot{z}$ of the line.

But,

But, by the propoſition, the whole force on DFA muſt be a minimum, or the fluent of $\frac{\dot{y}^3}{\dot{x}^2 + \dot{y}^2}$ muſt be a minimum when that of $\dot{x}$ becomes equal to the conſtant quantity DC; in which caſe it will be found that $\frac{\dot{x}\dot{y}^3}{\overline{\dot{x}^2 + \dot{y}^2}|^2}$ muſt be always equal to a conſtant quantity q; and hence $\dot{x}\dot{y}^3 = q \times \overline{\dot{x}^2 + \dot{y}^2}|^2$.

Now in this equation it is evident that $\dot{x}$ is to $\dot{y}$ in a conſtant ratio; but if two fluxions be always in a conſtant ratio, their fluents x, y, are known to be alſo in a conſtant ratio, which is the property of a right line.

Wherefore DFA is a right line, and the end ADB of the pier muſt be a right-lined triangle, that the force of the water upon it may be the leaſt poſſible.

PROPOSITION XII.

To determine the reſiſtance of the end of a pier againſt the ſtream of water.

Solution.

USING here the figure and notation of the laſt propoſition, by the ſame it is found that the fluxion of the force of the ſtream againſt the face DF is $\frac{\dot{y}^3}{\dot{x}^2 + \dot{y}^2}$; and ſince the fluxion of the force againſt the baſe is $\dot{y}$, it follows that the force of the ſtream againſt the baſe AB is to the force againſt the face ADB, as (y) the fluent of $\dot{y}$ is to the fluent of $\frac{\dot{y}^3}{\dot{x}^2 + \dot{y}^2}$. That is, the the abſolute force of the ſtream is to the efficacious force againſt the face of the pier, as its breadth is to double the fluent of $\frac{\dot{y}^3}{\dot{x}^2 + \dot{y}^2}$ when y is equal to half the breadth.

Corollary 1.

IF the face ADB be rectilineal.

Putting DC = a, CA = b, and AD = ($\sqrt{aa + bb} =$) c; as $a : b :: \dot{x} : \dot{y}$ by ſimilar triangles;

triangles; hence $x = \frac{ay}{b}$, and $\dot{x} = \frac{a\dot{y}}{b}$; this being written for it in the general expression above, we have $\frac{\dot{y}^3}{\frac{a^2\dot{y}^2}{b^2} + \dot{y}^2} = \frac{bb\dot{y}}{aa+bb} = \frac{bb\dot{y}}{cc}$ for the fluxion of the force on AD; the fluent of which, or $\frac{bby}{cc}$, is the force itself. And consequently the force on the flat base AB is to that on the triangular end, as y to $\frac{bby}{cc}$, or as cc to bb, that is, as AD^2 to AC^2.

And if AC be equal to CD, or ADB a right angle, which is generally the case, then $AD^2 = 2AC^2$, and the force on the base to that on the face, as 2 to 1.

Moreover, as the force on ADB, when ADB is a right angle, is only half of the absolute force, so it is evident that the force will be more than one-half when ADB is greater than a right angle, and less when it is less; and also that the longer AD is, the less the force is, it being always inversely as the square of AD.

Corollary 2.

IF ADB be a femicircle.

The radius AC = CD = a; then $2ax - xx = yy$, or $x = a - \sqrt{aa - yy}$, and $\dot{x} = \frac{y\dot{y}}{\sqrt{aa-yy}}$; hence $\frac{\dot{y}^3}{\dot{x}^2 + \dot{y}^2}$ becomes $\frac{\dot{y}^3}{\frac{y^2\dot{y}^2}{aa-yy} + \dot{y}^2} = \frac{aa-yy}{aa} \times \dot{y}$, the fluent of which is $\frac{aa - \frac{1}{3}yy}{aa} \times y$; and therefore the force on the bafe is to the force on the circular end, as y is to $\frac{aa - \frac{1}{3}yy}{aa} \times y$, or as aa to $aa - \frac{1}{3}yy$, or as $3aa$ to $3aa - yy$.

And when $y = a =$ AC, the proportion becomes that of 3 to 2.

So that only one-third of the abfolute force is taken off by making the end a femicircle.

Corollary 3.

WHEN the face ADB is a parabola.

Then, the notation being as before, viz. DC $= a$, and AC $= b$, we have $a : bb :: x : yy$; hence $x = \frac{ayy}{bb}$, and $\dot{x} = \frac{2ay\dot{y}}{bb}$; which being written

A TABLE of the natural RISE of WATER, in Proportion to the Refiftance or Obftruction it meets with, in its Paffage.

Conftruction of a modern bridge of 2 arches.	Velocity of the Current in one Second	OBSTRUCTIONS, OR RESISTANCES.							Stages of Accumulation in Floods	Conftruction of an ancient bridge of 3 or more arches.
Refiftance 1-11th		1-8th	1-4th	3-8ths	1-half	5-8ths	3-4ths	7-8ths		Refiftance 5-18ths
Rife of Water		Proportional Rife of Water in Feet, Inches, and Parts.								Rife of Water
F. I. Pts.		F. I. Pts.	F. I. Pts.	F. I. Pts.	F. I. Pts.	F. I. Pts.	F. I. Pts.	F. I. Pts.		F. I. Pts.
0 0 ·133	1 foot	0 0 ·158	0 0 ·283	0 0 ·49	0 0 ·87	0 1 ·69	0 4 ·041	1 4 ·728	Uniform Tenors.	0 0 ·320
0 0 ·533	2 feet	0 0 ·635	0 1 ·133	0 7 ·96	0 3 ·48	0 6 ·77	1 4 ·164	5 6 ·9		0 1 ·28
0 1 ·2	3 feet	0 1 ·428	0 2 ·549	0 4 ·41	0 7 ·835	1 3 ·234	3 0 ·368	12 6 ·53	Ordinary Floods.	0 2 ·881
0 2 ·133	4 feet	0 2 ·539	0 5 ·439	0 7 ·89	1 1 ·928	2 3 ·08	5 4 ·656	22 3 ·6		0 5 ·119
0 3 ·333	5 feet	0 3 ·967	0 7 ·083	1 0 ·25	1 9 ·763	3 6 ·316	8 5 ·024	34 10 ·31	Extraordinary Floods.	0 8 ·003
0 4 ·799	6 feet	0 5 ·713	0 10 ·199	1 5 ·64	2 7 ·339	5 0 ·934	12 1 ·476	50 2 ·112		0 11 ·525
Pier 12 / River 132	Velocities above feldom happen.	Piers 20 / Arches 140	Piers 40 / Arches 140	Piers 60 / Arches 160	Piers 80 / Arches 160	Piers 100 / Arches 160	Piers 120 / Arches 160	Piers 140 / Arches 160	Torrents above generally Inundations.	Piers 50 / River 180

This, next to one arch, which has no refiftance, without the flood encroaches on its crown, the moft eligible mode.

N. B. Thefe feveral numbers, refpectively, fhew how high the water is conftrained to rife above its natural level, or furface; which would otherwife, carry it off, in a free and uninterrupted paffage; therefore thefe numbers muft everywhere be added to the depth of water, below the fall, to give the true height of the flood.——The feven predicaments above fhew the excellence or imperfection of bridges, of every conftruction, and in all ftates of a flood, either in its uniform or variable tenors; and by which appears the great advantage of bridges of a fufficient capacity, and the pernicious confequences of all fuch as are not fo.——London bridge is nearly in the 6th predicament of this table, and Weftminfter bridge nearly in the 2nd. At the 1ft of thefe the Thames, with a velocity of about 3f. 2 in. per fecond, rifes to about 4f. 7in. and at the latter, with a velocity of 2·5 f. per fecond, to only 2·5 inches.

In this moft common mode feldom fufficient in a flood, the water foon encroaches on the arches, and changes the predicament.

This table to face page 77.

The material originally positioned here is too large for reproduction in this reissue. A PDF can be downloaded from the web address given on page iv of this book, by clicking on 'Resources Available'.

written in the general expreſſion, the fluent of it becomes the circular arc whoſe radius is $\frac{bb}{2a}$ and tangent y; and ſo the abſolute force is to the force on the parabolic end, as y to the arc whoſe tangent is y and radius $\frac{bb}{2a}$; that is, as the tangent of an arc is to the arc itſelf, the radius being to the tangent as 2 to $\frac{bb}{ay}$. And when $y = b$, the ratio of the tangent to radius, is that of 2 to $\frac{b}{a}$; or that of 2 to 1 when $DC = CA$. In which caſe the whole force is to the force on the parabolic end, as the tangent is to the arc of which the tangent is double the radius; that is, as the tangent of 63° 26′ 4″ to the arc of the ſame, or as 2 to 1·10714; which is a leſs force than on the circle, but greater than on the triangle.

And ſo on for other curves; in which it will be found that the nearer they approach to right lines, the leſs the force will be, and that it is leaſt of all in the triangle, in which it is one-half of the whole abſolute force when right-angled.

The annexed folding-out ſheet ſhews at one view the riſe of the water under the arches ariſing from its obſtruction by the piers, according to ſeveral rates of velocity, &c.

SEC-

SECTION V.

Of the Terms or Names of the various parts peculiar to a Bridge, and the Machines, &c. uſed about it; diſpoſed in alphabetical order.

ABUTMENT, or BUTMENT, which ſee in its place below.

ARCH, an opening of a bridge, through or under which the water, &c. paſſes, and which is ſupported by piers or by butments.

Arches are denominated circular, elliptical, cycloidal, catenarian, &c. according to the figure of the curve of them. There are alſo other denominations of circular arches according to the different parts of a circle: So, a ſemicircular arch is half the circle; a ſcheme or ſkeen arch is a ſegment leſs than the ſemicircle; and arches of the third and fourth point, or gothic arches, conſiſt of two circular arcs, excentric and meeting in an angle at top, each being 1-3d or 1-4th, &c. of the whole circle.

The chief properties of the moſt conſiderable arches, with regard to the extrados they require, &c. may be learned from the ſecond ſection. It there appears that none, but the arch of of equilibration in the example to prop. 5, can admit of a horizontal line at top; that this arch is not only of a graceful but of a convenient form, as it may be made higher or lower at pleasure with the same opening; that it, but no other, with a horizontal top, can be equally strong in all its parts, and therefore ought to be used in all works of much consequence. All the other arches require tops that are curved either upward or downward, some more and some less: Of these the elliptical arch seems to be the fittest to be substituted instead of the equilibrial one with any tolerable degree of propriety; it is in general also the best form for most bridges, as it can be made of any height to the same span, or of any span to the same height, while at the same time its hanches are sufficiently elevated above the water, even when it is pretty flat at top; which is a property of which the other curves are not possessed in the same degree; and this property is the more valuable, because it is remarked that after an arch is built and the centering struck, it settles more about the hanches than the other parts, by which other curves are reduced near to a streight line at the hanches. Elliptical arches also look bolder, are really stronger, and require less materials and labour than the others. Of the other curves, the cycloidal arch is next in quality to the elliptical one, for all the above properties. And, lastly, the circle. As to the others, the para-

parabola, hyperbola, and catenary, they may not at all be admitted in bridges of ſeveral arches; but may in ſome caſes be uſed for a bridge of one ſingle arch which is to riſe very high, becauſe then not much loaded at the hanches. We may hence alſo perceive the falſity of thoſe arguments which aſſert, that becauſe the catenarian curve ſupports itſelf equally in all its parts, it will therefore beſt ſupport any additional weight laid upon it: for the additional building made to raiſe the bridge to a horizontal line, or nearly ſuch, by preſſing more in one part than another, muſt force thoſe parts down, and the whole muſt fall. Whereas other curves will not ſupport themſelves at all without ſome additional parts built above them, to balance them, or to reduce their parts to an equilibrium.

ARCHIVOLT, the curve or line formed by the upper ſides of the vouſſoirs or arch ſtones. It is parallel to the intrados or underſide of the arch when the vouſſoirs are all of the ſame length; otherwiſe not.

By the archivolt is alſo ſometimes underſtood the whole ſet of vouſſoirs.

BANQUET, the raiſed foot path at the ſides of the bridge next the parapet. This ought to be allowed in all bridges of any conſiderable ſize:

ſize : it ſhould be raiſed about a foot above the middle or horſe paſſage, made 3, 4, 5, 6, 7, &c. feet broad according to the ſize of the bridge, and paved with large ſtones whoſe length is equal to the breadth of the walk.

BATTARDEAU, or *Coffer-dam*, a caſe of piling, &c. without a bottom, fixed in the bed of the river, water-tight or nearly ſo, by which to lay the bottom dry for a ſpace large enough to build the pier on. When it is fixed, its ſides reaching above the level of the water, the water is pumped out of it, or drawn off by engines, &c. till the ſpace be dry; and it is kept ſo by the ſame means, if there are leaks which cannot be ſtopped, till the pier is built up in it; and then the materials of it are drawn up again.

Battardeaux are made in various manners, either by a ſingle incloſure, or by a double one, with clay or chalk rammed in between the two, to prevent the water from coming through the ſides. And theſe incloſures are alſo made either with piles only, driven cloſe by one another, and ſometimes notched or dove-tailed into each other; or with piles grooved in the ſides, driven in at a diſtance from one another, and boards let down between them in the grooves.

M The

The method of building in battardeaux cannot well be uſed where the river is either deep or rapid. It alſo requires a very good natural bottom of ſolid earth or clay; for, although the ſides be made water-tight, if the bottom or bed of the river be of a looſe conſiſtence, the water will ooze up through it in too great abundance to be evacuated by the engines.

It is almoſt needleſs to remark that the ſides muſt be made very ſtrong, and well propt or braced in the inſide, to prevent the ambient water from preſſing the ſides in, and forcing its way into the battardeau.

BRIDGE, a work of carpentry or maſonry, built over a river, canal, &c. for the conveniency of croſſing the ſame.

A ſtone bridge is an edifice forming a way over a river, &c. ſupported by one arch or by ſeveral arches, and theſe again ſupported by proper piers or butments.

A ſtately bridge over a large river is one of the moſt noble and ſtriking pieces of art. To behold huge and bold arches, compoſed of an immenſe quantity of ſmall materials, as ſtones, bricks, &c. ſo diſpoſed and united together that

they

they ſeem to form but one ſolid compact body, affording a ſafe paſſage for men and carriages over large waters, which with their navigation paſs free and eaſy under them at the ſame time, is a ſight truly ſurprizing and affecting indeed.

To the abſolutely neceſſary parts of a bridge already mentioned, viz. the arches, piers, and abutments, may be added the paving at top, the parapet wall, either with or without a baluſtrade, &c. alſo the banquet or raiſed foot way on each ſide, leaving a ſufficient breadth in the middle for horſes and carriages. The breadth of a bridge for a great city ſhould be ſuch as to allow an eaſy paſſage for three carriages and two horſemen a-breaſt in the middle way, and for three foot paſſengers in the ſame manner on each banquet. And for other leſs bridges a leſs breadth.

As a bridge is made for a way or paſſage over a river, &c. ſo it ought to be made of ſuch a height as will be quite convenient for that paſſage; but yet ſo as to be conſiſtent with the intereſt and concerns of the river itſelf, eaſily admitting through its arches the craft that navigate upon it, and all the water even at high tides and floods. The neglect of this precept has been the ruin of many bridges, and particularly that at Newcaſtle, over the river Tyne, on the 17th of november 1771. So that in determining its height, the conveniencies both of the paſſage over it and under it ſhould be conſidered, and the height made to anſwer the beſt for them both, obſerving to make the *convenient* give place to the *neceſſary* when their intereſts are oppoſite.

Bridges are generally placed in a direction perpendicular to the ſtream in a direct line, to give free paſſage to the water, &c. But ſome think they ſhould be made not in a ſtreight line, but convex towards the ſtream, the better to reſiſt floods, &c. And ſome ſuch bridges have been made.

Again, a bridge ſhould not be made in too narrow a part of a navigable river, or one ſubject to tides or floods: becauſe the breadth being ſtill more contracted by the piers, will increaſe the depth, velocity, and fall of the water under the arches, and endanger the whole bridge and navigation.

The number of arches of a bridge are generally made odd; either that the middle of the ſtream or chief current may flow freely without the interruption of a pier; or that the two halves of the bridge, by gradually riſing fro the ends to the middle, may there meet in the higheſt and largeſt arch; or elſe, for the ſake of grace, that by being open in the middle, the eye in viewing

viewing it may look directly through there, as one always expects to do in looking at it, and without which opening one generally feels a disappointment in viewing it.

If the bridge be equally high throughout, the arches, being all of a height, are made all of a size; which causes a great saving of centering. If the bridge be higher in the middle than at the ends, let the arches decrease from the middle towards each end, but so as that each half have the arches exactly alike, and that they decrease in span, proportionally to their height, so as to be always the same kind of figure, and similar parts of that figure: thus, if one be a semicircle, let the rest be semicircles also, but proportionally less; if one be a segment of a circle, let the rest be similar segments of other circles; and so for other figures. The arches being equal at equal distances on both sides of the middle, is not only for the strength and beauty of the bridge, but that the centering of the one half may serve for the other also. But if the bridge be higher at the ends than in the middle, the arches ought to increase in span and pitch from the middle towards the ends.

When the middle and ends are of different heights, their difference however ought not to be great in proportion to the length, that the ascent may be easy; and then also it is more beautiful

beautiful to make the top one continued curve than two inclined ſtreight lines from the ends towards the middle.

Bridges ſhould rather be of few and large arches than of many and ſmall ones, if the height and ſituation will poſſibly allow of it; for this will leave more free paſſage for the water and navigation, and be a great ſaving in materials and labour, as there will be fewer piers and centers, and the arches themſelves will require leſs materials.

For the fabric of a bridge, and the proper eſtimation of the expence, &c. there are generally neceſſary three plans, three ſections, and an elevation. The three plans are ſo many horizontal ſections, viz. the firſt a plan of the foundation under the piers, with the particular circumſtances attending it, whether of gratings, planks, piles, &c. the ſecond is the plan of the piers and arches, &c. and the third is the plan of the ſuperſtructure, with the paved road and banquet. The three ſections are vertical ones; the firſt of them a longitudinal ſection from end to end and through the middle of the breadth; the ſecond a tranſverſe one, or acroſs it, and through the ſummit of an arch; and the third alſo acroſs, and taken upon a pier. The elevation is an orthographic projection of one ſide or face of the bridge, or its appearance as viewed

at

at an infinite diſtance, and ſhews the exterior aſpect of the materials, and the manner in which they are worked and decorated.

Other obſervations are to be ſeen in the firſt ſection.

BUTMENTS, or *abutments*, the extremities of a bridge, by which it joins to or abuts upon the land or ſides of the river, &c. Theſe muſt be made very ſecure, quite immovable, and more than barely ſufficient to reſiſt the drift of its adjacent arch. So that if there are not rocks or very ſolid banks to raiſe them againſt, they muſt be well reinforced with proper walls or returns, &c. The thickneſs of them that will be barely ſufficient to reſiſt the ſhoot of the arch, may be calculated as that of a pier by prop. 10.

When the foundation of a butment is raiſed againſt a ſloping bank of rock, gravel, or good ſolid earth, it will produce a ſaving of materials and labour, to carry the work on by returns at different heights, like ſteps of ſtairs.

CAISSON, a kind of *cheſt*, or flat-bottomed boat, in which a pier is built, then ſunk to the bed of the river, and the ſides looſened and taken off from the bottom, by a contrivance for that purpoſe; the bottom of it being left under

under the pier as a foundation. It is evident therefore that the bottoms of caiſſons muſt be made very ſtrong and fit for the foundations of the piers. The caiſſon is kept a-float till the pier be built to about the height of low-water mark; and for that purpoſe its ſides muſt either be made of more than that height at firſt, or elſe gradually raiſed to it as it ſinks by the weight of the work, ſo as always to keep its top above water. And therefore the ſides muſt be made very ſtrong, and kept aſunder by croſs timbers within, leſt the great preſſure of the ambient water cruſh the ſides in, and ſo not only endanger the work, but alſo drown the men who work within it. The caiſſon is made of the ſhape of the pier, but ſome feet wider on every ſide to make room for the men to work: the whole of the ſides are of two pieces, both joined to the bottom quite around, and to each other at the ſalient angle, ſo as to be diſengaged from the bottom and from each other when the pier is raiſed to the deſired height, and ſunk. It is alſo convenient to have a little ſluice made in the bottom, occaſionally to open and ſhut, to ſink the caiſſon and pier ſometimes by, before it be finiſhed, to try if it bottom level and rightly; for by opening the ſluice, the water will ruſh in and fill it to the height of the exterior water, and the weight of the work already built will ſink it; then by ſhutting the ſluice again, and pumping out the water,

water, it will be made to float again, and the rest of the work may be completed: but it must not be sunk but when the sides are high enough to reach above the surface of the water, otherwise it cannot be raised and laid dry again. Mr. Labelye tells us that the caissons in which he built some of the piers of Westminster bridge, contained above 150 load of fir timber of 40 cubic feet each, and was of more tonnage or capacity than a 40 gun ship of war.

CENTERS, are the timber frames erected in the spaces of the arches to turn them on, by building on them the voussoirs of the arch. As the center serves as a foundation for the arch to be built on, when the arch is completed, that foundation is struck from under it, to make way for the water and navigation, and then the arch will stand of itself from its curved figure. A center must therefore be constructed of the exact figure of the intended arch, convex as the arch is concave, to receive it on as a mould. If the form be circular, the curve is struck from a central point by a radius: if it be elliptical, it ought to be struck with a doubled cord, passing over two pins or nails fixed in the focusses, as the mathematicians describe their ellipses; and not by striking different pieces or arcs of circles from several centers; for these will form no ellipse at all, but an irregular misshapen curve made up of broken pieces of different

ferent circular arcs: but if the arch be of any other form, the ſeveral abſciſſas and ordinates ought to be calculated, then their correſponding lengths, transferred to the centering, will give ſo many points of the curve, and exactly by which points bending a bow of pliable matter, the curve may be drawn by it.

The centers are conſtructed of beams, &c. of timber firmly pinned and bound together, into one entire compact frame, covered ſmooth at top with planks or boards to place the vouſſoirs on, the whole ſupported by offsets in the ſides of the piers, and by piles driven into the bed of the river, and capable of being raiſed and depreſſed by wedges, contrived for that purpoſe, and for taking them down when the arch is completed. They ought alſo to be conſtructed of a ſtrength more than ſufficient to bear the weight of the arch.

In taking the center down; firſt let it down a little, all in a piece, by eaſing ſome of the wedges; there let it reſt a few hours or days to try if the arch make any efforts to fall, or any joints open, or ſtones cruſh or crack, &c. that the damage may be repaired before the center is entirely removed, which is not to be done till the arch ceaſes to make any viſible efforts.

In ſome bridges the centering makes a very conſiderable part of the expence, and therefore all means of ſaving in this article ought to be cloſely attended to; ſuch as making few arches, and as nearly alike or ſimilar as poſſible, that the centering of one arch may ſerve for others, and at leaſt that the ſame center may be uſed for both of each pair of equal arches on both ſides of the middle.

CHEST, the ſame as *Caiſſon.*

COFFERDAM, the ſame as *Battardeau.*

DRIFT, *Shoot*, or *Thruſt* of an arch, is the puſh or force which it exerts in the direction of the length of the bridge. This force ariſes from the perpendicular gravitation of the ſtones of the arch, which, being kept from deſcending by the form of the arch and the reſiſtance of the pier, exert their force in a lateral or horizontal direction. This force is computed in prop. 10, where the thickneſs of the pier is determined that is neceſſary to reſiſt it; and is greater the lower the arch is, *cæteris paribus.*

ELEVATION, the orthographic projection of the front of a bridge on the vertical plane parallel to its length. This is neceſſary to ſhew the form and dimenſions of the arches and other parts as to height and breadth, and therefore has a plain ſcale annexed to it to meaſure the parts by. It alſo ſhews the manner of working up and decorating the fronts of the bridge.

EXTRADOS, the exterior curvature or line of an arch. In the propoſitions of the ſecond ſection it is the outer or upper line of the wall above the arch; but it often means only the upper or exterior curve of the vouſſoirs.

FOUNDATIONS, the bottoms of the piers, &c. or the baſes on which they are built. Theſe bottoms are always to be made with projections, greater or leſs according to the ſpaces on which they are built. And according to the nature of the ground, depth and velocity of water, &c. the foundations are laid and the piers built after different manners, either in caiſſons, in battardeaux, on ſtilts with ſterlings, &c. for the particular methods of doing which, ſee each under its reſpective term.

The moſt obvious and ſimple method of laying the foundations and raiſing the piers up to water-mark, is to turn the river out of its courſe above the place of the bridge, into a new channel cut for it near the place where it makes an elbow or turn; then the piers are built on dry ground, and the water turned into its old courſe again, the new one being ſecurely banked up.

up. This is certainly the best method, when the new channel can be easily and conveniently made; but which however is seldom or never the case.

Another method is to lay only the space of each pier dry till it be built, by surrounding it with piles and planks driven down into the bed of the river, so close together as to exclude the water from coming in; then the water is pumped out of the inclosed space, the pier built in it, and lastly the piles and planks drawn up. This is coffer-dam work, but evidently cannot be practised if the bottom be of a loose consistence admitting the water to ooze and spring up through it.

When neither the whole nor part of the river can be easily laid dry as above, other methods are to be used; such as to build either in caissons or on stilts, both which methods are described under their proper words; or yet by another method, which hath, though seldom, been sometimes used, without laying the bottom dry, and which is thus: the pier is built upon strong rafts or gratings of timber well bound together, and buoyed up on the surface of the water by strong cables, fixed to other flotes or machines, till the pier is built; the whole is then gently let down to the bottom, which must be made level for the purpose. But of these

these methods, that of building in caissons is the best.

But before the pier can be built in any manner, the ground at the bottom must be well secured, and made quite good and safe if it be not so naturally. The space must be bored into to try the consistence of the ground; and if a good bottom of stone, or firm gravel, clay, &c. be met with within a moderate depth below the bed of the river, the loose sand, &c. must be removed and digged out to it, and the foundation laid on the firm bottom on a strong grating or base of timber made much broader every way than the pier, that there may be the greater base to press on, to prevent its being sunk. But if a solid bottom cannot be found at a convenient depth to dig to, the space must then be driven full of strong piles, whose tops must be sawed off level some feet below the bed of the water, the sand having been previously digged out for that purpose; and then the foundation on a grating of timber laid on their tops as before. Or, when the bottom is not good, if it be made level, and a strong grating of timber, two, three, or four times as large as the base of the pier be made, it will form a good base to build on, its great size preventing it from sinking. In driving the piles, begin at the middle, and proceed outwards all the way to the borders or margin: the reason of which is, that if if the outer ones were driven firſt, the earth of the inner ſpace would be thereby ſo jammed together, as not to allow the inner piles to be driven. And beſides the piles immediately under the piers, it is alſo very prudent to drive in a ſingle, double, or triple row of them around and cloſe to the frame of the foundation, cutting them off a little above it, to ſecure it from ſlipping aſide out of its place, and to bind the ground under the pier the firmer. For, as the ſafety of the whole bridge depends on the foundation, too much care cannot be uſed to have the bottom made quite ſecure.

JETTEE, the border made around the ſtilts under a pier, being the ſame with *Sterling*.

IMPOST, is the part of the pier on which the feet of the arches ſtand, or from which they ſpring.

KEYSTONE, the middle vouſſoir, or the arch ſtone in the top or immediately over the center of the arch. The length of the keyſtone, or thickneſs of the archivolt at top, is allowed to be about 1-15th or 1-16th of the ſpan, by the beſt architects.

ORTHOGRAPHY, the elevation of a bridge, or front view as ſeen at an infinite diſtance.

PARAPET,

PARAPET, the breaſt wall made on the top of a bridge to prevent paſſengers from falling over. In good bridges, to build the parapet but a little part of its height cloſe or ſolid, and upon that a baluſtrade to above a man's height, has an elegant effect.

PIERS, the walls built for the ſupport of the arches, and from which they ſpring as their baſes.

They ought to be built of large blocks of ſtone, ſolid throughout, and cramped together with iron, which will make the whole as one ſolid ſtone. Their faces or ends, from the baſe up to high-water mark, ought to project ſharp out with a ſalient angle, to divide the ſtream. Or, perhaps, the bottom of the pier ſhould be built flat or ſquare up to about half the height of low-water mark, to allow a lodgment againſt it for the ſand and mud, to cover the foundation; leſt, by being left bare, the water ſhould in time undermine and ſo ruin or injure it. The beſt form of the projection for dividing the ſtream, is the triangle; and the longer it is, or the more acute the ſalient angle, the better it will divide it, and the leſs will the force of the water be againſt the pier; but it may be ſufficient to make that angle a right one, as it will make the work ſtronger, and in that caſe the per-

perpendicular projection will be equal to half the breadth or thickneſs of the pier. In rivers on which large heavy craft navigate and paſs the arches, it may perhaps be better to make the ends ſemicircular ; for although it does not divide the water ſo well as the triangle, it will both better turn off and bear the ſhock of the craft.

The thickneſs of the piers ought to be ſuch as will make them of weight or ſtrength ſufficient to ſupport their interjacent arch independent of any other arches. And then if the middle of the pier be run up to its full height, the centering may be ſtruck to be uſed in another arch before the hanches are filled up. The whole theory of the piers may be ſeen in the third ſection.

They ought to be made with a broad bottom on the foundation, and gradually diminiſhed in thickneſs by offsets up to low-water mark.

The methods of laying their foundations, and building them up to the ſurface of the water, are given under the word FOUNDATION.

PILES, are timbers driven into the bed of the river for various purpoſes, and are either round, ſquare, or flat like planks. They may be of any wood which will not rot under water, but oak and fir are moſtly uſed, eſpecially the latter, on account of its length, ſtreightneſs, and cheapneſs. They are ſhod with a pointed iron at the bottom, the better to penetrate into the ground; and are bound with a ſtrong iron band or ring at top, to prevent them from being ſplit by the violent ſtrokes of the ram by which they are driven down.

Piles are either uſed to build the foundations on, or are driven about the pier as a border of defence, or to ſupport the centers on; and in this caſe. when the centering is removed, they muſt either be drawn up or ſawed off very low under water; but it is perhaps better to ſaw them off and leave them ſticking in the bottom, leſt the drawing of them out ſhould looſen the ground about the foundation of the pier. Thoſe to build on, are either ſuch as are cut off by the bottom of the water, or rather a few feet within the bed of the river; or elſe ſuch as are cut off at low-water mark, and then they are called ſtilts. Thoſe to form borders of defence, are rows driven in cloſe by the frame of a foundation, to keep it firm; or elſe they are to form a caſe or jettee about ſtilts, to keep within it the ſtones that are thrown in to fill it up; in this caſe, the piles are grooved, driven at a little diſtance from each other, and plank piles let into the grooves between them, and driven down alſo, till the whole ſpace is ſurrounded. Beſides uſing this for ſtilts, it is alſo ſometimes neceſſary to ſurround a ſtone pier with

with a ſterling or jettee, and fill it up with ſtones to ſecure an injured pier from being ſtill more damaged, and the whole bridge ruined. The piles to ſupport the centers may alſo ſerve as a border of piling to ſecure the foundation, cutting them off low enough after the center is removed.

PILE DRIVER, an engine for driving down the piles. It conſiſts of a large ram of iron ſliding perpendicularly down between two guide poſts; which being lift up to the top of them, and there let fall from a great height, comes down upon the top of the pile with a violent blow. It is worked either with men or horſes, and either with or without wheel work. That which was uſed at the building of Weſtminſter bridge, is perhaps the beſt ever invented.

PITCH, of an arch, the perpendicular height from the ſpring or impoſt to the keyſtone.

PLAN, of any part, as of the foundations, or piers, or ſuperſtructure, is the orthographic projection of it on a plane parallel to the horizon.

PUSH, of an arch, the ſame as drift, ſhoot, &c.

SALIENT ANGLE, of a pier, the projection of the end againſt the ſtream, to divide it. The right-lined angle beſt divides the ſtream, and the more acute the better for that purpoſe; but the right angle is generally uſed as making the beſt maſonry. A ſemicircular end, though it does not divide the ſtream ſo well, is ſometimes better in large navigable rivers, as it carries the craft the better off, or bears their ſhocks the better.

SHOOT, of an arch, the ſame as drift.

SPRINGERS, are the firſt or loweſt ſtones of an arch, being thoſe at its feet bearing immediately on the impoſt.

STERLINGS, or Jettees, a kind of caſe made about a pier of ſtilts, &c. to ſecure it, and is particularly deſcribed under the next word *Stilts*.

STILTS, a ſet of piles driven into the ſpace intended for the pier, whoſe tops being ſawed level off about low-water mark, the pier is then raiſed on them. This method was formerly uſed when the bottom of the river could not be laid dry; and theſe ſtilts were ſurrounded, at a few feet diſtance, by a row of piles and planks, &c. cloſe to them like a coffer-dam, and called a ſterling or jettee; after which looſe ſtones,

ſtones, &c. are thrown or poured down into the ſpace till it be filled up to the top, by that means forming a kind of pier of rubble or looſe work, and which is kept together by the ſides or ſterlings: this is then paved level at the top, and the arches turned upon it. This method was formerly much uſed, moſt of the large old bridges in England being erected that way, ſuch as London bridge, Newcaſtle bridge, Rocheſter bridge, &c. But the inconveniencies attending it are ſo great, that it is now quite exploded and diſuſed: for, becauſe of the looſe compoſition of the piers, they muſt be made very large or broad, or elſe the arch would puſh them over and ruſh down as ſoon as the center was drawn; which great breadth of piers and ſterlings ſo much contracts the paſſage of the water, as not only very much incommodes the navigation through the arch, from the fall and quick motion of the water, but from the ſame cauſe alſo the bridge itſelf is in much danger, eſpecially in time of floods, when the water is too much for the paſſage. Add to this that beſides the danger there is of the pier burſting out the ſterlings, they are alſo ſubject to much decay and damage by the velocity of the water and the craft paſſing through the arches.

THRUST, the ſame as drift, &c.

VOUS-

VOUSSOIRS, the ſtones which immediately form the arch, their under ſides conſtituting the intrados. The middle one, or keyſtone, ought to be about 1-15th or 1-16th of the ſpan, as has been obſerved; and the reſt ſhould increaſe in ſize all the way down to the impoſt; the more they increaſe the better, as they will the better bear the great weight which reſts upon them without being cruſhed, and alſo will bind the firmer together. Their joints ſhould alſo be cut perpendicular to the curve of the intrados.

THE END

Printed in the United States
By Bookmasters